Storm-petrels of the Eastern Pacific Ocean: Species Assembly and Diversity along Marine Habitat Gradients

ORNITHOLOGICAL MONOGRAPHS

Editor: John Faaborg

224 Tucker Hall
Division of Biological Sciences
University of Missouri
Columbia, Missouri 65211

Project Manager: Mark C. Penrose
Managing Editor: Richard D. Earles

AOU Publications Office
622 Science Engineering
Department of Biological Sciences
University of Arkansas
Fayetteville, Arkansas 72701

The *Ornithological Monographs* series, published by the American Ornithologists' Union, has been established for major papers and presentations too long for inclusion in the Union's journal, *The Auk*. All material in this monograph may be copied for noncommercial purposes of educational or scientific advancement without need to seek permission.

Copies of *Ornithological Monographs* are available from Buteo Books, 3130 Laurel Road, Shipman, VA 22971. Price of *Ornithological Monographs* no. 62: $10.00 ($9.00 for AOU members). Add $4.00 for handling and shipping charges in U.S., and $5.00 for shipping to other countries. Make checks payable to Buteo Books.

Authors of this issue: Larry B. Spear and David G. Ainley

Library of Congress Control Number 2006938295

Printed by Cadmus Communications, Lancaster, PA 17601

Issued 15 January 2007

Ornithological Monographs, No. 62 xii + 77 pp.

ISBN: 0-943610-71-0

Cover: (Top left) White-faced Storm-Petrel (*Pelagodroma marina*), southwest Pacific, off New Zealand (S. N. G. Howell); (top right) Band-rumped Storm-Petrel (*Oceanodroma castro*), South Atlantic, near St. Helena (S. N. G. Howell); (bottom left) Ashy Storm-Petrel (*O. homochroa*), California Current, off Monterey (S. N. G. Howell); (bottom right) Wilson's Storm-Petrel (*Oceanites oceanicus*), Scotia Sea (D. G. Ainley).

Translation of abstract by Rafael Brito-Aguilar.

Storm-petrels of the Eastern Pacific Ocean: Species Assembly and Diversity along Marine Habitat Gradients

BY

LARRY B. SPEAR AND DAVID G. AINLEY

H. T. Harvey and Associates, 3150 Almaden Expressway, Suite 145, San Jose, California 95118, USA

ORNITHOLOGICAL MONOGRAPHS NO. 62

PUBLISHED BY
THE AMERICAN ORNITHOLOGISTS' UNION
WASHINGTON, D.C.
2007

TABLE OF CONTENTS

LIST OF TABLES

APPENDICES:

LIST OF FIGURES

From the Editor

I was born and raised in Iowa, which is about as far from the ocean as one can get in North America. On the other hand, because my home town was situated on the drift plain from fairly recent glaciation, I became very comfortable living with ocean-scale flatness covered with corn and beans as far as the eyes could see. Later in life, when I started studying habitat fragmentation in the Midwest, it was easy to adopt the analogy of "islands of habitat" in a "sea" of agriculture.

Although it seems easy to think of the ocean as a flat, uniform habitat (as we might for the vast corn-growing region of the Midwest), my ecological training and life experiences told me otherwise. Dissertation work in the West Indies took me throughout the Caribbean, where the water always seemed warm, even in winter when the locals felt it was much too cold for swimming. During a project in the Galápagos Islands, I learned what a cold ocean feels like, even when exposed to the equatorial sun.

So this Iowa boy quickly figured out that the ocean is a very heterogeneous habitat, with places where the surface waters vary between cold and warm and places where currents stir nutrients to the surface (and places where they don't). Of course, through the powers of flight, birds have evolved ways to take advantage of the resources found on the ocean, even when the distances involved and the navigational abilities required seem overwhelming. Most of these true seabirds spend the bulk of the year far from any shoreline, though recent work suggests that these movements are anything but "wandering," as some of the common names of these birds might suggest. The one great constraint that they all face is that they must return to land to nest, and islands that provide safe nesting sites are quite limited in comparison with the vastness of potential foraging areas.

To those of us who still believe in community ecology and its associated concepts of assembly rules and ecological isolating mechanisms, this truly oceanic avifauna is a vast unknown. Combine the occurrence of dozens of seabird species ranging in size from sparrow-sized storm-petrels to enormous albatrosses with an ocean that offers habitats differing in the amounts and types of food available, and numerous classic ecological questions jump to mind. Do similar-sized species divide this ocean into pieces? Do coexisting species differ by size or diet in some regular fashion? Can we explain the number of coexisting seabird species in different locations by looking at oceanic productivity in those sites? How does the diversity of the vast Pacific Ocean avifauna compare to that of smaller oceans across the globe? Does the fact that these species must go to land to breed affect any orderly patterns that may occur during the nonbreeding season? Alternatively, given that these species can travel thousands of miles across the ocean, is it just a confusing mixture of birds across the ocean, as species ignore what we see as ocean structure and truly wander around? One could go on and on with possibilities to explore and hypotheses to test.

This monograph starts to answer some of these questions by looking at distributional patterns in the storm-petrels of the eastern Pacific. These are the tiniest of seabirds, ranging from sparrows to robins in size despite living on the open ocean, and they are surprisingly diverse in this region. The authors have spent incredible amounts of time identifying seabirds while crossing the ocean; much of the monograph is the first documentation of the at-sea ranges of many of these birds and the first descriptions of much of their nonbreeding ecology. After describing so much new material about each of these species, the authors synthesize their results by looking for patterns of distribution and coexistence among this set of seabirds.

Of course, I think every reader should read every word of this monograph, and I am sure the authors agree. On the other hand, if your time is limited and you want to get the big picture of what is going on, you might skip through many of the species accounts after you have looked at one or two. By reading the introductory material, a species account or two, and the final synthetic material, you will get a great introduction to seabird community ecology. It might stimulate you to go back and read some more accounts, or perhaps it will encourage you to jump on a boat and cross the Pacific, counting seabirds along the way. This Iowa boy won't join you, because he gets seasick really easily.

John Faaborg

Larry B. Spear, 1945–2006
(Photograph courtesy of S. N. G. Howell.)

Ornithological Monographs
Volume (2007), No. 62, 1–77

Printed in USA.

STORM-PETRELS OF THE EASTERN PACIFIC OCEAN: SPECIES ASSEMBLY AND DIVERSITY ALONG MARINE HABITAT GRADIENTS

Larry B. Spear and David G. Ainley[1]

H. T. Harvey and Associates, 3150 Almaden Expressway, Suite 145, San Jose, California 95118, USA

Abstract.—Residing in waters just west of the Americas, in a roughly triangular area from about 50°N to 50°S and out along the equator, are 26 distinct taxa of storm-petrels (Hydrobatidae), a diversity far greater for this group than in any equivalent stretch of the world ocean. We sought to understand how so many forms can co-occur within this very poorly known family.

We describe the ranges, at-sea behavior, and marine habitat affinities of 22 (of the 26) distinct forms included within 5 (of 6) genera and 16 (of 20) species of storm-petrel, all of which occur in that portion of the Pacific Ocean that stretches from the California Current south through the Humboldt Current and out to 170°W along the equator. We base our analysis on data collected during 23 cruises conducted in the eastern Pacific during the period 1980–1995. We also provide, for the first time for most forms, information on the annual cycle, as well as abundance estimates based on at-sea censusing during both the breeding and nonbreeding periods. Such information is unknown for almost all populations of storm-petrels, unless their at-sea range has been thoroughly censused. We include the following storm-petrels in our analysis: Leach's (*Oceanodroma leucorhoa*; represented by a light-rumped form, *O. l. leucorhoa*, and several dark-rumped forms: *O. l. socorroensis*, *O. l. cheimomnestes*, and *O. l. chapmani*), Band-rumped (*O. castro*), Ashy (*O. homochroa*), Least (*O. microsoma*), Wedge-rumped (*O. tethys*; represented by two races: *O. t. tethys* and *O. t. kelsalli*), Black (*O. melania*), Markham's (*O. markhami*), Ringed (*O. hornbyi*), Wilson's (*Oceanites oceanicus*; represented by two races: *O. o. oceanicus* and *O. o. chilensis*), White-vented (*Oceanites gracilis*; represented by two races: *O. g. gracilis* and *O. g. galapagoensis*), White-bellied (*Fregetta grallaria*; represented by three races: *F. g. grallaria*, *F. g. segethi*, and *F. g. titan*), Black-bellied (*F. tropica*), White-throated (*Nesofregetta fuliginosa*), and White-faced (*Pelagodroma marina*; represented by two races: *P. m. dulciae* and *P. m. maoriana*).

Information was gathered by strip censuses (400–600 m wide), observations of storm-petrel behavior along cruise tracks, and collection of specimens. Within the entire study area, we made 9,308.1 h of observation and surveyed 111,029 km^2 of ocean, including 61,131 km^2 in boreal spring–austral autumn and 49,898 km^2 in boreal autumn–austral spring. Surveys included 768.3 h within 1,000 km of the South American coast; >11,203.7 km^2 of ocean was surveyed, 7,382.1 km^2 in austral autumn and 3,821.6 km^2 in austral spring. Surveys within 500 km of North America included 2,557.2 h over 12,473 km^2 in boreal spring and 3,061.3 km^2 in boreal autumn. We also collected specimens during numerous stops where oceanographic studies were being conducted by other researchers who were also aboard the ship.

For the majority of taxa, our surveys covered the entire at-sea range of the taxon. We had complete coverage for the following storm-petrels: Ringed, White-vented, Markham's, *O. t. segethi* race of White-bellied, both races of Wedge-rumped, Galápagos race of Band-rumped (*O. c. bangsi*), races of dark-rumped Leach's (*O. l. socorroensis*, *O. l. cheimomnestes*, and *O. l. chapmani*), and Ashy. During boreal autumn, we also had nearly complete coverage for Black and Least storm-petrels, both of which vacate the Gulf of California after the breeding season. We also had nearly complete coverage of the Pacific range of the White-throated Storm-Petrel. Using generalized additive models, population estimation was quite satisfactory for these taxa.

Our results indicate that most storm-petrel taxa in the study area have robust populations, this report presenting the first estimates ever for most of the taxa treated. On the other hand, meager populations are indicated for Ashy and White-throated storm-petrels, for two races of White-bellied Storm-petrel (*F. g. grallaria* and *F. g. titan*), and for Band-rumped Storm-Petrels in Hawaii. All appear to have populations of <10,000 birds, especially in the case of the newly

[1]E-mail: dainley@penguinscience.com

rediscovered (present study) but apparently nearly extinct population of *F. g. titan* on Rapa Island.

The eastern Pacific is oceanographically heterogeneous at the middle to large scale, and such heterogeneity with strong environmental gradients apparently contributes in a major way to the diversity of storm-petrels in the area. The occurrence patterns of all forms sorted along gradients, such as those for sea-surface temperature and salinity and thermocline depth and strength, all of which separate the major current systems and water masses in the region. Except for the three endemic storm-petrels of the Humboldt Current (White-vented, Ringed, and Markham's storm-petrels), species were further sorted at a smaller scale in accord with a habitat gradient from shelf to slope to pelagic waters. Gradients in ocean productivity correspond with, and are affected by, gradients in the above physical habitat features. Flight and foraging behavior also differentiated species ecologically. Finally, intense competition for nesting space is indicated for many taxa by clear evidence of "floating populations" (i.e., surpluses of breeding adults denied the chance to breed owing to lack of nesting space). In part, these surpluses are the result of intense competition for limited nesting space and are resolved by body size, with a size differential dictating what taxa can breed sympatrically (on the same island). At sea, especially in areas of high ocean productivity where habitats are more finely demarcated than elsewhere, storm-petrels sort by foraging habitat likely derived from trophic competition that was far more intense during glacial periods. At that time, continental shelves were much narrower and, therefore, depth-defined foraging zones were more closely packed than at present or throughout the Holocene. Where species co-occur at sea, the two factors that explain co-occurrence among most storm-petrel species are foraging habitat and body size, with competition for food being resolved by the latter, which affects prey size, as well as by foraging behavior. Overall, the high degree of speciation among storm-petrels of the eastern Pacific is likely a product of relatively sparse nesting islands sprinkled among a confluence of distinct water types separated by intense environmental and productivity gradients. *Received 15 June 2005, accepted 19 September 2006.*

Resumen.—Existen 26 taxa diferentes de paiños (Hydrobatidae) en un área casi triangular en los 50° N y 50° S y a lo largo del ecuador en las aguas al oeste de América, lo que representa una diversidad mucho mayor para este grupo en comparación con cualquier otra zona de los océanos mundiales. Buscamos entender como es que tantas formas diferentes pueden presentarse dentro de esta familia tan poco conocida.

Describimos las variedades, en comportamiento en altamar, y las afinidades de hábitats marinos de 22 (de las 26) formas distintas que están incluidas dentro de los 5 (de los 6) géneros y 16 (de las 20) especies de paíños, los cuales todos se presentan en esa parte del Océano Pacifico que se angosta desde el sur de la corriente de California hasta la corriente de Humboldt y hasta los 170° oeste a lo largo del Ecuador. Apoyamos nuestro análisis en datos obtenidos durante 23 cruceros en el este del Pacífico durante el periodo de 1980-1995. También ofrecemos, por primera vez para la mayoría de las formas, información en el ciclo anual, así como también estimaciones de abundancia basadas en censos en altamar durante la temporada de reproducción y fuera de ésta. Tal información es desconocida para casi todas las poblaciones de paíños, a menos que su rango en altamar haya sido estudiado a fondo. Incluimos las siguientes especies de paíños en nuestro análisis: *Oceanodroma leucorhoa* (representada por una forma de rabadilla clara, *O. l. leucorhoa*, y varias formas de rabadilla obscura, *O. l. socorroensis*, *O. l. cheimomnestes*, y *O. l. chapmani*), *O. castro*, *O. homochroa*, *O. microsoma*, *O. tethys* (representada por dos razas: *O. t. tethys* y *O. t. kelsalli*), *O. melania*, *O. markhami*, *O. hornbyi*, *Oceanites oceanicus* (representada por dos razas: *O. o. oceanicus* and *O. o. chilensis*), *O. gracilis* (representada por dos razas: *O. g. gracilis* and *O. g. galapagoensis*), *Fregetta grallaria* (representada por tres razas: *F. g. grallaria*, *F. g. segethi*, and *F. g. titan*), *F. tropica*, *Nesofregetta fuliginosa*, y *Pelagodroma marina* (representada por dos razas: *P. m. dulciae* and *P. m. maoriana*).

La información se obtuvo mediante censos en franjas (400–600 m de ancho), observaciones del comportamiento de paíños a lo largo de la travesía de los cruceros y la colecta de especímenes. En toda el área de estudio, realizamos 9,308.1 h de observaciones y muestreamos 111,029 km^2 de océano, incluyendo 61,131 km^2 en primavera boreal-otoño austral y 49, 898 km^2 en otoño boreal-primavera austral. Los muestreos incluyeron 768.3 h en 1,000 km a lo largo de la costa de Sudamérica; mas de 11,203.7 km^2 de océano fue muestreado, 7,382.1 km^2 en otoño austral y 3,821.6 km^2 en primavera austral. Los muestreos dentro de 500 km de Norte América incluyeron 2,557.2 h sobre 12,473 km^2 en primavera boreal y 3,061.3 km^2 en otoño boreal. También colectamos especímenes durante numerosas paradas dónde se realizaron estudios oceanográficos por otros investigadores quienes también estaban a bordo del barco.

Para la mayoría de las taxa, nuestros muestreos cubrieron el rango total en altamar para el taxón. Tuvimos una cobertura completa para las siguientes especies: *O. hornbyi*; *O. gracilis*; *O. markhami*; *O. t. segethi*; *O. t. tethys* and *O. t. kelsalli*; *O. c. bangsi*; *O. l. socorroensis*, *O. l. cheimomnestes*, *O. l. chapmani* y *O. homochroa*. Durante el otoño austral, también tuvimos una cobertura casi completa para *O. melania* y *O. microsoma*, las cuales desocupan el Golfo de California después de la temporada de reproducción. También tuvimos una cobertura casi completa en la distribución del pacifico de *N. fulginosa*. La estimación poblacional usando modelos aditivos generalizados fue bastante satisfactoria para estos taxa.

Nuestros resultados indican que la mayoría de los taxa de paíños en el área de estudio tienen poblaciones robustas, este reporte presenta las primeras estimaciones para la mayoría de los taxa estudiados. Por otro lado, las poblaciones pequeñas están representadas por *O. homochroa*; *N. fulginosa*; *F. g. grallaria* and *F. g. titan*; y *O. castro* en Hawai. Todas estas especies parecen tener poblaciones mayores de 10,000 aves, especialmente en el caso de la población recientemente redescubierta (este estudio) pero aparentemente casi extinta de *F. g. titan* en la isla Rapa.

El este del océano pacifico es, oceanográficamente, homogéneo a una escala mediana a larga, y esta heterogeneidad con gradientes ambientales aparentemente fuertes contribuye grandemente a la diversidad de paíños en el área. Los patrones de presencia de todas las formas a través de los gradientes, tales como aquellos de temperatura superficial del mar y salinidad y profundidad y fuerza de las termoclinas, las cuales separan la mayoría de los sistemas actuales y masas de aguas en la región. A excepción de tres especies endémicas de paíños de la corriente de Humboldt (*O. gracilis*, *O. hornbyi*, and *O. markhami*), las especies fueron clasificadas a una escala mas pequeña de acuerdo con un gradiente de hábitat desde la plataforma a la pendiente hasta las aguas pelágicas. Los gradientes en productividad oceánica corresponden con, y son afectados por, gradientes en las características físicas del hábitat. El vuelo y el comportamiento de forrajeo también diferenciaron a las especies ecológicamente. Finalmente, la intensa competencia por espacio para anidar esta indicado para muchos taxa por medio de la clara evidencia de "poblaciones flotantes" (los excesos de adultos en reproducción que no pudieron aparearse debido ala falta de espacio para anidar). En parte, estos excesos de individuos son el resultado de la intensa competencia por espacio limitado para anidar y se resuelve por el tamaño del cuerpo, con una diferencia de tamaño que dicta que taxa puede reproducirse simpátricamente (en la misma isla). En alta mar, especialmente en áreas de alta productividad oceánica dónde hábitats están mas finamente demarcados que en ninguna otra parte, paíños clasificados por sus hábitos de forrajeo probablemente se derivaron de competencias tróficas que fueron mucho mas intensas durante los periodos glaciares. Durante esta era las plataformas continentales estaban mas angostas y, por lo tanto, las zonas de forrajeo definidas por profundidad estaban mas estrechas que en el presente o a través del Holoceno. Donde las especies concurren en altamar, los dos factores que explican esta concurrencia entre la mayoría de las especies de paíños son hábitat de forrajeo y tamaño corporal, siendo la competencia por alimento resuelta por esta ultima, lo que afecta el tamaño de la presa así como también por el comportamiento de forrajeo. En general, el alto grado de especiación entre los paíños del este del Pacifico es probablemente producto de islas de anidación relativamente esparcidas entre distintos tipos de aguas separadas por intensos gradientes de productividad y ambientales.

Introduction

The family of storm-petrels (Hydrobatidae) is not as speciose as other families in the order Procellariiformes (albatrosses, shearwaters, fulmarine petrels, gadfly petrels, prions, and diving-petrels) but is equivalent in diversity to the albatrosses (Diomedeidae) and shearwaters (Puffininae). Yet, heretofore, it has been among the least known of avian families. Storm-petrels are the most pelagic of birds and among the most widespread, distributed across two-thirds of the Earth's surface, including all oceans (Brooke 2004). In spite of such a wide distribution, they are rarely encountered by humans, because most people, including ornithologists, rarely venture very far from shore. Storm-petrels are unique, being the smallest of all seabirds, ranging in size from that of a sparrow (e.g., *Zonotrichia* spp.) up to that of an American Robin (*Turdus migratorius*).

Because storm-petrels are strictly confined to the open ocean and come to land only to breed, they have long been shrouded in mystery and myth. They were known to mariners as "Mother Carey's chickens," in reference to the witch who possessed the key to "Davy Jones's Locker," where drowned sailors were said to repose. Acknowledging the storm-petrels' behavior of using their broad, webbed feet to push off from the sea surface to change direction and, perhaps, their tendency to collect in the lee of ships during fierce storms, mariners considered the word "petrel" a reference to St. Peter, who was said to have walked on water (Lockley 1983). Although they have been noted and studied at sea for hundreds of years, information on these birds' behavior and ecology has been very slow to accumulate, no doubt because their habitat is so foreign to most humans. A suite of antipredator strategies also obscures their life cycle: they come to land (mainly islands) only to nest and only at night, and reside there in narrow crevices, or burrows; they are quick to desert nests if only slightly disturbed by competitors, predators, or researchers (Warham 1990, Ambagis 2004).

The diversity of storm-petrel species is particularly noteworthy in the eastern Pacific Ocean. Indeed, 6 genera, 17 species, and, among the latter, 26 distinct races are found in the waters stretching along the American coast within 50°N and 50°S of the equator out to Hawaii (Murphy 1936, Crossin 1974, Marchant and Higgins 1990). These include, with nomenclature hereafter following Warham (1990), the following storm-petrels: Leach's (*Oceanodroma leucorhoa*; represented by a light-rumped form, *O. l. leucorhoa*, and several dark-rumped forms: *O. l. socorroensis*, *O. l. cheimomnestes*, and *O. l. chapmani*), Band-rumped (*O. castro*), Fork-tailed (*O. furcata*), Sooty (*O. tristrami*), Ashy (*O. homochroa*), Least (*O. microsoma*), Wedge-rumped (*O. tethys*; represented by two races: *O. t. tethys* and *O. t. kelsalli*), Black (*O. melania*), Markham's (*O. markhami*), Ringed (*O. hornbyi*), Wilson's (*Oceanites oceanicus*; represented by two races: *O. o. oceanicus* and *O. o. chilensis*), White-vented (*O. gracilis*; represented by two races: *O. g. gracilis* and *O. g. galapagoensis*), White-bellied (*Fregetta grallaria*; represented by three races: *F. g. grallaria*, *F. g. segethi*, and *F. g. titan*), Black-bellied (*F. tropica*), White-throated (*Nesofregetta fuliginosa*), Gray-backed (*Garrodia neiris*), and White-faced (*Pelagodroma marina*; represented by two races: *P. m. dulciae* and *P. m. maoriana*). The only storm-petrel species lacking in the eastern Pacific is the British Storm-Petrel (*Hydrobates pelagicus*). Were it not for predation by feral cats at the nesting grounds, the Guadalupe Storm-Petrel (*O. macrodactyla*) would be included.

At-sea distributions of most of these species are known broadly (Harrison 1983, and references above), but their fine- and mesoscale occurrence patterns and the factors that affect their discontinuous presence within their range are poorly known. Moreover, information on the localities, abundance, and annual cycle of most storm-petrel species is grossly incomplete. In fact, nests of three of the eastern Pacific species have yet to be found (see below), which is not very surprising, given that storm-petrels are so difficult to study on their breeding grounds, as noted above. Because of their secretive nesting habits, estimating their abundance is also especially problematic, and the estimates that have been done are rough, at best (e.g., Ainley 1995, Huntington et al. 1996, Ainley and Everett 2001, Boersma and Silva 2001). Yet knowledge of population size and trends is required for assessing and monitoring these species' status in this modern world of fishery depletion, widespread pollution, and introduction of alien species to seabird nesting islands—all factors that have severely affected storm-petrels (see references just above) and other seabirds. Finally, owing to the logistic difficulties inherent in at-sea study, little work has been conducted to describe the at-sea behavior, habitat use, and pelagic ecology of these species, in the eastern Pacific or elsewhere, or factors responsible for their extensive taxonomic radiation, even among species groups occupying the same oceanographic systems.

Indeed, why there are so many storm-petrel taxa in the eastern Pacific, a diversity several times greater than that of any other seabird group in any comparable expanse of ocean, has never been addressed. Owing to the subductance of oceanic plates beneath the western portion of American continents, there are relatively few islands for breeding in this area. On the other hand, the two eastern boundary currents, California and Humboldt, and the equatorial upwelling along the interface of the North and South Pacific gyres, provide some of the most productive marine waters anywhere on Earth and a great deal of environmental heterogeneity, with several distinct water types separated by remarkable environmental gradients (Longhurst and Pauly 1987, Ballance et al. 1997, Spear et al. 2001; Fig. 1). Species abundance and community composition in the avifauna are known to respond to these factors (e.g., Ribic and Ainley 1988; Ballance et al. 1997; Oedekoven et al. 2001; Spear et al. 1995a, 2001, 2003; Spear and Ainley 2005a; Vilchis et al. 2006). Environmental gradients are also involved, no doubt, in the radiation evident among hydrobatids in the eastern Pacific (a concept reviewed in Mayr 1964, Pianka 1966, and Faaborg 1988). Therefore, we began our studies using the hypothesis that the diversity of storm-petrel taxa in this region reflected the marine habitat complexity. Testing that hypothesis, however, required a lot of data collection on the basic population status and habitat relationships of species in this group (see below), subjects that constitute the bulk of this monograph. In retrospect, we also amassed enough information

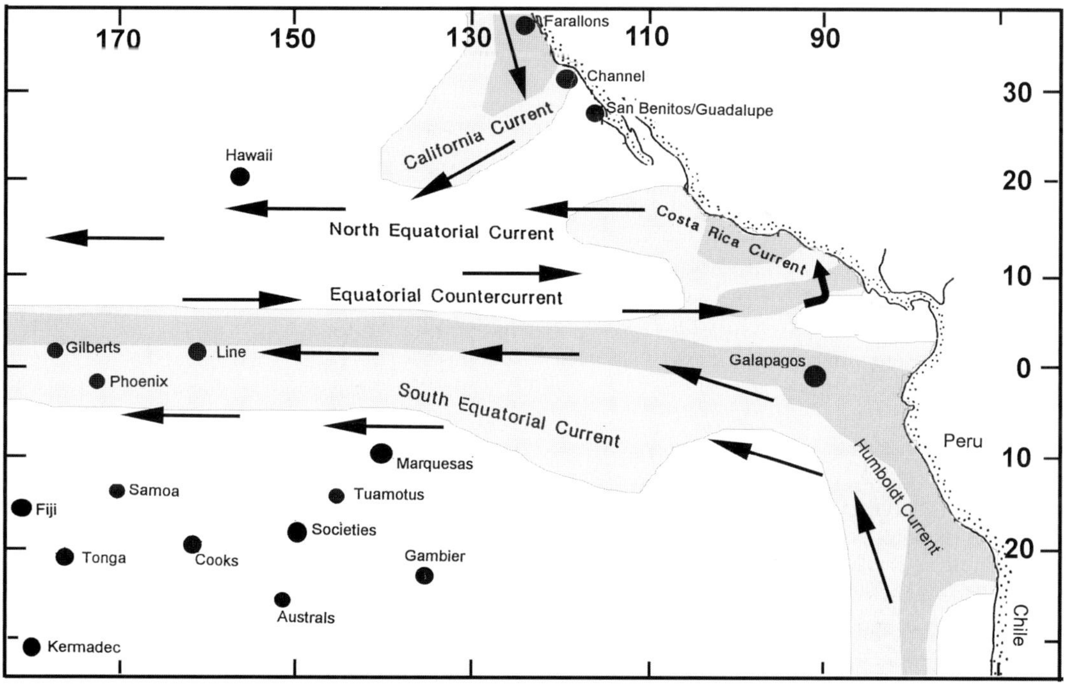

FIG. 1. Major current systems and storm-petrel breeding islands discussed in the text, superimposed on the large-scale patterns of ocean productivity. The three gradations of productivity are <200, 201–300, and >300 mgC m^{-2} day^{-1}. In large part, these patterns were taken from Longhurst and Pauly (1987), who did not include in their analysis the ocean north of 20°N (i.e., the region of the California Current). The pattern for the latter area in this figure is an approximation based on images of ocean color (chlorophyll concentration being related strongly, but not 1:1, to ocean productivity). The region between 20°N and 20°S is referred to in the text as "eastern tropical Pacific."

on species' population sizes and range overlap, as well as trophic patterns and relationships (Spear et al. 2007), that we have been able to propose "rules" under which eastern Pacific storm-petrels may or may not co-occur (e.g., Diamond 1975, Brown 1981, Faaborg 1985).

Herein, we report findings based on at-sea surveys conducted over 20 years in the eastern Pacific Ocean. Besides insights into ocean habitat use, behavior, and estimates of abundance (Clarke et al. 2003), our at-sea studies were designed to provide data on the relative contribution of subadults and adults to the total population (e.g., Spear et al. 2003, Spear and Ainley 2005a). Indeed, abundance of nonbreeders is an important indicator of the status of the entire population (Klomp and Furness 1992). To that end, and to provide information on the adult population component, we also examined ratios of adults to subadults among samples of each species that we collected in the eastern Pacific Ocean. Elsewhere, complementing the present paper, we report on the diet and feeding ecology of storm-petrels compared with other seabirds of the eastern Pacific (Spear et al. 2007).

Review of Existing Information

We saw no Sooty, Black-bellied, or Gray-backed storm-petrels, which occur only at the margin of our study area, and we covered very little of the range of Fork-tailed Storm-Petrel. Therefore, we have excluded these four species from the present study, which includes analyses for 22 morphologically distinct taxa of storm-petrels, as summarized below.

Leach's Storm-Petrel.—The light-rumped form of Leach's Storm-Petrel is one of the few somewhat well-known members of the family, thanks to studies mainly in waters off eastern Canada (Huntington et al. 1996). In the eastern Pacific, it breeds during boreal spring and summer (with the exception of one race; Crossin 1974, Ainley 1980) on islands stretching along the Pacific rim from Russia to Baja California and, during the nonbreeding period, ranges at sea into the southern Pacific Ocean (Brooke 2004), sometimes well to the south. Indeed, individuals apparently prospecting for breeding sites have been caught on the Chatham Islands near New Zealand (Imber and Lovegrove 1982), another was found dead on a beach near Auckland (Falla 1933), and several have been found on beaches in Australia (Marchant and Higgins 1990). Its molt does not overlap the breeding season (Ainley et al. 1976). Several races have been described for the Pacific, though some are not readily separable either in the field or mensurally (Crossin 1974, Power and Ainley 1986). Estimates of breeding populations of light-rumped Leach's Storm-Petrels include eastern Russia, 350,000 (Brooke 2004); Japan, 2 million (Hasegawa 1984); Alaska, 4 million (Lensink 1984); and British Columbia to California, 800,000 individuals (Brooke 2004).

We considered a Leach's Storm-Petrel "dark-rumped" if one-half or more of the upper tail coverts were dark-colored as opposed to white (see Ainley 1980). Dark-rumped Leach's Storm-Petrels breed during boreal spring and summer from the Farallon Islands south to the San Benito Islands off Baja California in a cline of decreasing prevalence among light-rumped forms (Ainley 1980) and occur at sea from about 45°N to 10°S, generally within several hundred kilometers of the coast of the Americas (Crossin 1974, present study). The breeding population estimate for *O. l. chapmani* (entirely dark-rumped form) on the Coronodos and San Benito islands is 50,000, whereas the breeding population estimate for *O. l. socorroensis* and *O. l. cheimomnestes* (mixed color morphs) on Guadalupe Island is ~10,000 birds (Everett and Anderson 1991).

Recently, Vilchis et al. (2006) found that the distribution of Leach's Storm-Petrel (type not specified) in the eastern tropical Pacific (which is a part of our eastern Pacific Ocean study area) responded to chlorophyll concentration, sea-surface temperature, and thermocline characteristics. They reasoned that these factors must be proxies for the plankton on which they presumed these birds fed.

Band-rumped Storm-Petrel.—Populations of this species breed during boreal spring and summer on the Hawaiian Islands (*O. c. cryptoleucura*; Harrison et al. 1984) and islands off Japan (race unknown; Cheke 1967, Hasegawa 1984). The nesting behavior of this form is virtually unknown. However, on the Galápagos Islands, these storm-petrels (race *O. c. bangsi*; Murphy 1936, Harris 1969) have two breeding seasons that are six months out-of-phase: one during austral winter (May to July) and the other in austral summer, from November to February (Harris 1969). Individuals breeding during the two periods do so consistently from year to year. Races of Band-rumped Storm-petrels known from the Pacific are morphologically inseparable (Austin 1952). This species has been observed at sea primarily within ~1,000 km of the Galápagos Islands (Crossin 1974, Spear and Ainley 1999), though small numbers have been observed south of Hawaii to the equator (Crossin 1974) and a single bird was recorded off San Diego, California (McCaskie 1990). The breeding population of ~10 colonies on the Galápagos Islands was estimated at 30,000 birds (Harris 1969, Coulter 1984), that of Hawaii is unknown (nests have not been found; Harrison et al. 1984), and that of Japan is estimated at 50,000 (Cheke 1967, Enticott and Tipling 1997).

Least and Black storm-petrels.—These species breed during boreal summer on islands off the Pacific coast of Baja and southern California (Crossin 1974, Jehl 1984) and within the Gulf of California (Velarde and Anderson 1994). At sea, they generally occur within several hundred kilometers of the Americas, from central California to Peru (Murphy 1936, Crossin 1974, Ainley 1976). Both species have large populations breeding on the San Benito Islands off the Pacific coast of Baja California (Brooke 2004), though the majority of both species breed on islands in the Gulf of California (Velarde and Anderson 1994). The total population of Least Storm-Petrels is believed to number in the hundreds of thousands, and possibly millions (Everett and Anderson 1991). The population of Black Storm-Petrels breeding on the San Benito Islands may exceed 1 million (Ainley and Everett and 2001). Although the number breeding in the Gulf of California is unknown, they may exceed 500,000 (Brooke 2004).

Ashy Storm-Petrel.—This species is endemic to the California Current and breeds during boreal spring and summer, mainly on the Farallon and Channel islands off the coast of California. At sea, this species occurs abundantly from ~38.5°N (Bodega Bay) to ~36.3°N (Point Piños on southern Monterey), and sparsely from there to the Pacific coast of Baja California (Crossin 1974, present study). We are not aware of records north of 38.5°N or south of ~32°N. Records by Crossin (1974) from 47°N to 7°N and out to 480 km offshore likely represented misidentified dark-rumped Leach's Storm-Petrels. The Ashy Storm-Petrel's breeding biology is fairly well known on the basis of studies at the Farallon Islands (Ainley et al. 1990, Ainley 1995); its season is greatly extended during the year, and its molt broadly overlaps breeding, unlike the more migratory Leach's Storm-Petrels, which also breed at the Farallons (Ainley et al. 1976).

We had light survey coverage of waters south of Pt. Piños, especially those of the Southern California Bight, where the second important breeding location of this species occurs (Channel Islands). Therefore, we confine our analyses for the Ashy Storm-Petrel to the area we surveyed extensively, from 38.5°N to 36.5°N, which includes waters from Bodega Head to Point Piños (i.e., those adjacent to the Farallon Islands, the most important breeding location of this species). In 1992, using capture–recapture methods on Southeast Farallon Island, Sydeman et al. (1998), estimated a population of 4,284, consisting of both breeding and nonbreeding individuals. This estimate, however, is probably low, because netting was conducted only in several locations on that island and did not include any netting on other of the Farallon Islands where this species likely nests.

Wedge-rumped Storm-Petrel.—The nominate race of Wedge-rumped Storm-Petrel, *O. t. tethys*, is known to breed only on the Galápagos Islands, with an estimated 400,000 birds breeding on Tower (Genovesa) Island (Harris 1969). Numbers breeding on other islands in the archipelago are unknown (Coulter 1984). These birds have an extended breeding season that appears to be centered in austral spring and summer (Harris 1969, Crossin 1974). *Oceanodroma t. kelsalli* breeds during austral winter on the Pacific coast of South America, where the only known breeding locations are the San Gallan, Pescadores, La Vieja, and San Lorenzo islands off Peru (Murphy 1936; see also Duffy et al. 1984). Breeding population size of *kelsalli* has not been estimated.

The two races of Wedge-rumped Storm-Petrels can be distinguished mensurally on the basis of wing chord, which does not overlap between the two; wing chord is shorter in *kelsalli* (Murphy 1936). Together, these storm-petrels range north regularly to Baja California, west on the equator to ~150°W, and south to central Chile (Murphy 1936, Crossin 1974, Vilchis et al. 2006). Most Wedge-rumped Storm-Petrels collected within 500 km of Mexico, including Baja California, have been *kelsalli*, and most collected beyond 500 km of shore have been *tethys* (Moffitt 1938, Huey 1952, Crossin 1974, present study). Crossin (1974) also reported on a series of Wedge-rumped Storm-Petrels collected in the Gulf of Panama, all of which were *kelsalli*.

White-vented, Ringed, and Markham's storm-petrels.—These species are endemic to the Humboldt Current. Two races of White-vented Storm-Petrels are known, including a South American form, *O. g. gracilis*, and a Galápagos Islands form, *O. g. galapagoensis* (Murphy 1936). No nests of *galapagoensis* have been found, and only one nest of *gracilis* has been located (on Chungungo Island, northern Chile, in August 1979; Schlatter and Marin 1983). The Ringed Storm-Petrel is believed to breed in the very arid deserts occurring on the mainland coast of South America, though nests have not been found (Murphy 1936). Colonies of Markham's Storm-Petrels were discovered in 1992 on the arid Paracas Peninsula of Peru (Jahncke 1993).

Enlarged gonads and an egg in the oviducts of White-vented Storm-Petrels collected in the Humboldt Current in May indicate that this species breeds in austral winter (Murphy 1936). Egg laying by Markham's occurs from late June to August (J. Jahncke pers. comm.), which indicates an austral winter and spring breeding season. On the other hand, 50 Ringed Storm-Petrels collected in May–July showed little gonad development, and all had completed or were undergoing primary molt. Three females collected in August also were in nonbreeding condition (Crossin 1974). However, the occurrence of grounded fledglings in Antofagasta (coastal desert), Chile, and Lima, Peru, in June and July, indicates an austral autumn-to-winter breeding season, March to July (Brooke 2004; but see below).

Pelagic ranges of all three species include waters adjacent to the coast of Ecuador, Peru, and Chile (Murphy 1936), though Markham's Storm-Petrels have been seen west of the Galápagos Islands on the equator (Crossin 1974), in the Panama Bight (Spear and Ainley 1999), and off Baja California (Pyle 1993).

Abundance of these species was heretofore unknown (Duffy et al. 1984, Schlatter 1984), though an estimated total of 1,444 Markham's Storm-Petrel nests were present on the Paracas Peninsula in 1992 (Jahncke 1993). Abundance of Markham's and White-vented storm-petrels were each estimated to (likely) exceed 30,000 (Brooke 2004). Population size of Ringed Storm-Petrels was estimated to include thousands, probably tens of thousands of birds (Enticott and Tipling 1997).

Wilson's Storm-Petrel.—Three races of Wilson's Storm-Petrel have been described. These include a small-sized form (*O. o. chilensis*) that breeds during austral summer in the Tierra del Fuego region, Chile (Murphy 1936), and two distinctly larger forms that also breed during austral summer, on sub-Antarctic islands (*O. o. oceanicus*), mainland Antarctica, and the Scotia Sea Islands (*O. o. exasperatus*; Marchant and Higgins 1990). Birds of Chilean origin have been recorded north to Ecuador (Murphy 1936). The population size of breeding Chilean Wilson's Storm-Petrels is unknown (Schlatter 1984).

Birds sighted in the central Pacific south of Hawaii are believed to be of Antarctic origin (Crossin 1974), as indicated by a northward migration through the Marshall Islands in the postbreeding period of austral autumn (Huber 1971). Wilson's Storm-Petrels have frequently been recorded off western North America. The breeding population of Wilson's Storm-Petrels from sub-Antarctic latitudes and the Antarctic mainland numbers in the millions (e.g., Murphy 1936, Croxall et al. 1984, Brooke 2004).

White-bellied Storm-Petrel.—Marchant and Higgins (1990) and Brooke (2004) recognized four races of this species: (1) *F. g. grallaria*, breeding on Admiralty, Lord Howe, Macauley, and Kermadec islands off Australia and New Zealand; (2) *F. g. leucogaster*, breeding on Tristan da Cunha, St. Paul, and possibly Amsterdam islands in the southern Atlantic and Indian oceans; (3) *F. g. titan*, breeding on Rapa of the Austral Islands in southeastern Polynesia; and (4) *F. g. segethi*, breeding on the Juan Fernandez and Desventuradas islands off central Chile. The available information indicates that the breeding period, including egg laying to fledging, occurs primarily in summer to early autumn (January–May; Murphy 1936, Marchant and Higgins 1990). On the other hand, there is evidence that the breeding period of *F. g. segethi* may be more prolonged, including egg laying in January and fledging in June (Brooke 1987), as indicated by chicks of various ages found in June (Murphy 1936).

Fregetta g. grallaria is represented by melanistic individuals (birds that have darkly feathered areas on the belly and flanks), a high proportion of which come from the Lord Howe population (Murphy and Snyder 1952, Bell et al. 1984). Kermadec Island birds include light and intermediate color morphs (Tennyson and Taylor 1990), and Juan Fernandez birds are all light morphs (present study).

The at-sea range of these races has been, heretofore, incompletely known (Marchant and Higgins 1990). *Fregetta g. segethi* is believed to be confined primarily to waters of the Humboldt Current off South America, though a bird likely to have been *segethi* was observed in the Gulf of Panama (Spear and Ainley 1999); *F. g. grallaria* ranges at sea to the north and east of the breeding grounds off Australia; and *F. g. leucogaster* occurs within the south Atlantic and, possibly, the Indian Ocean. The pelagic range of *F. g. titan* is unknown; a bird believed to be of this race was collected 450 km south of the Galápagos Islands in 1906 (Murphy 1936).

The Lord Howe breeding population of *F. g. grallaria* probably exceeds 1,000 pairs on offshore stacks and islets, but does not appear to breed on Lord Howe itself (Baker et al. 2002). On the Kermadec Islands, the breeding population was estimated at <2,000 pairs (Robertson and Bell 1984, Taylor 2000). No estimates are available for breeding populations of *F. g. titan* or *segethi* (Garnett 1984, Schlatter 1984), though *F. g. leucogaster* is believed to be represented by 100,000 to 1 million breeders (reviewed in Brooke 2004).

White-throated Storm-Petrel.—This species breeds on several island groups in the central equatorial Pacific and on Easter Island (Murphy 1936, Crossin 1974). Both white-bellied and melanistic morphs occur. On the Line Islands (Christmas Island), breeding occurs throughout the year but egg laying is concentrated in August to December (austral spring and summer; Schreiber and Ashmole 1970). Most at-sea sightings of this species have occurred in equatorial waters, but in two disjunct regions: one in the central Pacific and the other in the eastern Pacific (Crossin 1974). The breeding population of this species on Gambier, Line, and Phoenix islands was estimated at 4,000 to 40,000 birds (Garnett 1984), though Brooke (2004) suggests that the world population could be no more than 10,000 birds. The possibility that populations of this species breed on the Marquesas, Samoan, Fiji, and Vanuatu islands has not been confirmed.

White-faced Storm-Petrel.—Six races of this species have been described, including three that nest in the Pacific, and three that nest in the Atlantic (Murphy and Irving 1951, Brooke 2004). The Pacific races breed during austral spring and summer: *P. m. dulciae* on islands off Australia, *P. m. maoriana* on islands off New Zealand, and *P. m. albiclunis* (doubtfully distinct; Marchant and Higgins 1990) may breed on the Kermadec Islands. These three forms can possibly be distinguished on the basis of plumage pattern and degree of graduation in length of tail quills (Murphy and Irving 1951). However, *maoriana* may not be separable from the Atlantic race, *P. m. marina*, which breeds on the Tristan da Cunha group and Gough Island in the southeast Atlantic during austral spring and summer (Ryan et al. 1990, Brooke 2004). A second Atlantic race, *P. m. hypoleuca*, breeds in austral spring and summer on the Salvage and Canary islands in the northeast Atlantic (Snow and Perrins 1998, Campos and Granadeiro 1999), and the third, *P. m. eadesi*, breeds on the Cape Verdes islands (Brooke 2004).

Movements of the Atlantic races are not well documented (Brooke 2004) but are not known to occur in the Pacific. New Zealand *maoriana* are believed to disperse primarily to the eastern Pacific, on the basis of a bird ringed on the Chatham Islands and recovered 9° south of the Galápagos Islands (Imber 1984). Referring to occurrences of this race near the Galápagos Islands, along the coast of Peru, and along the equator (Murphy and Irving 1951, Lévêque et al. 1966, Crossin 1974), the latter author suggested that they follow an eastern route to the southern Humboldt Current, follow the current north to the equator, and then migrate southwest back to New Zealand. On the basis of ringing recoveries, the Australian *dulciae* are believed to move into the Indian Ocean (Brooke 2004).

Breeding population estimates of Pacific races are as follows: *dulciae*, approaching 1 million birds (Baker et al. 2002); *maoriana*, 2–4 million birds (reviewed in Brooke 2004); and *albiclunis*, <200 birds (Taylor 2000).

Study Area and Methods

For the present study, we define the eastern Pacific Ocean to include the tropical ocean within 25° of the equator, from the coast of the Americas to 170°W, plus the subtropical to temperate (sub-Arctic) waters that stretch poleward, 400 km offshore, along the coast of the Americas—the Humboldt Current flowing south to ~50°S (Paulik 1981) and the California Current north to 45°N (Figs. 1 and 2).

Oceanographic characteristics of the study area.—Circulation of surface waters in the eastern Pacific Ocean is dominated by six major current systems. These include the equatorial currents: the Equatorial Counter Current (ECC), North Equatorial Current (NEC), and South Equatorial Current (SEC); the Humboldt and California currents, two of the world's five eastern boundary currents (eastern boundary of major ocean gyres); and, between the latter, the Costa Rica Current (a branch of the ECC) (Fig. 1, from Wyrtki 1966).

The ECC, which flows west to east, varies seasonally in both strength and location. During the periods in which we surveyed birds in the eastern Pacific (late March through June, and October through December), the ECC is strong and lies between 4°N and 10°N (Wyrtki 1966). The NEC and SEC lie to its north and south, respectively, and flow east to west. The ECC extends to 25°N, and the SEC extends to 20°S, though the latter is strongest between 4°N and 10°S (Wyrtki 1966).

The California Current, defined here as waters from 20°N to 45°N and from the coast to ~350 km offshore, flows south along western North America; off Baja California, it swings westward to merge with the NEC. The Costa Rica Current, defined here as waters from 20°N to the equator, flows north along the Central American and Mexican coasts and also merges with the NEC.

The Humboldt Current, defined here as waters from 0° to 50°N, flows north along South America and merges into the SEC near the Galápagos Islands. From the equator to 20°S, the Humboldt Current extends from the coast to

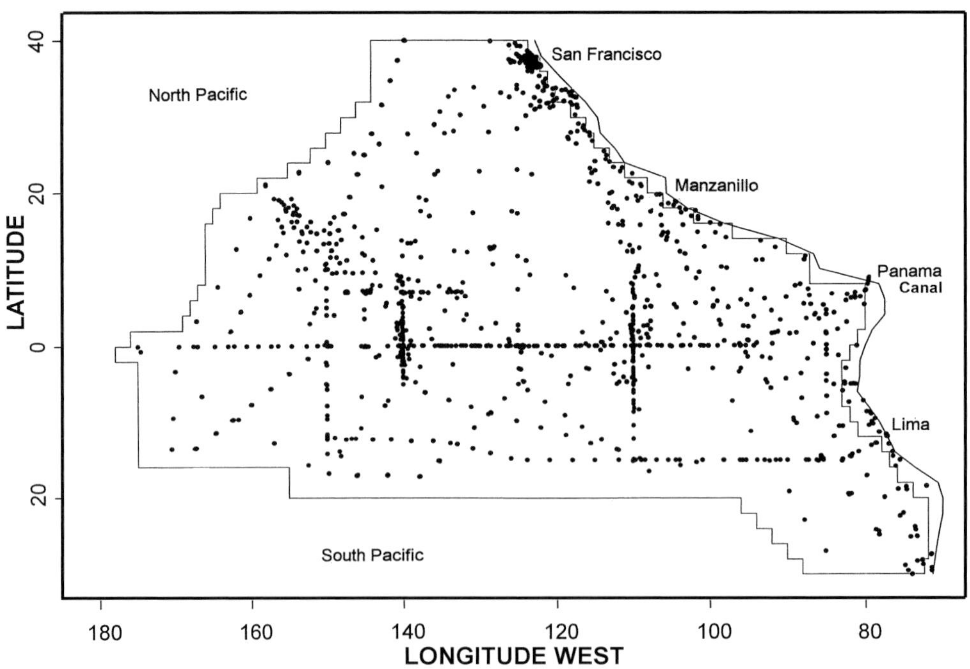

Fig. 2. Noon positions (n = 1,186) of ships on which we conducted seabird surveys in the eastern Pacific Ocean, 1980–1995. Positions north of 40°N and south of 30°S are not shown.

~450 km offshore; from 20°S to 40°S, it extends offshore only to ~300 km (Murphy 1936, Paulik 1981, Glantz and Thompson 1981). Maximum upwelling in the Humboldt Current occurs during austral winter, and may not vary appreciably with latitude, though evidence indicates that Chilean waters are separated from the Peruvian waters by a warm-water belt at ~20°S. A gradient in sea-surface salinity increases from south to north and from east to west (Wyrtki 1966). In general, sea-surface temperature shows less of a gradient south to north, compared with the strong temperature gradient inshore to offshore. Because of these facts, and to better describe storm-petrel distributions off South America, we divided the Humboldt Current according to species ranges in that current system. For Band-rumped, Markham's, Ringed, and White-vented storm-petrels, subdivisions included the northern Humboldt Current (0–20°S) and the central Humboldt Current (20–35°S); for Wilson's (*O. o. chilensis*), White-bellied (*F. g. segethi*), and White-faced storm-petrels, Humboldt Current subdivisions included approximately similar sections: 0–25°S and 25–55°S.

Of the three equatorial currents, the SEC has the lowest surface temperature and highest salinity; the ECC has the highest surface temperature and lowest salinity; and the NEC has intermediate values. Gradients in temperature, salinity, and thermocline depth and strength increase from east to west; by contrast, primary productivity, which can be very high where surface waters diverge along the Equatorial Front (boundary between NEC and ECC), increases from west to east. Productivity is highest east of 100°W at the equator, and east of 115°W on the northern border of the ECC (Fiedler et al. 1991; for effect on seabird foraging guilds, see Spear et al. 2001). The California and Humboldt currents both have cool, high-salinity waters with shallow, weak thermoclines and are among the most productive areas in the world as a result of wind-generated coastal upwelling (Glantz and Thompson 1981). The Costa Rica Current is warmer and less saline, with a variable thermocline.

Another important feature of the eastern Pacific Ocean is the Intertropical Convergence Zone. This area is characterized by high frequency of squalls, high winds, and heavy rainfall. It is often located at the northern boundary of the ECC (Wyrtki 1966).

The effect of El Niño on the distribution of seabirds in the eastern Pacific Ocean, including storm-petrels, has been examined by Ribic and Ainley (1997), Ribic et al. (1997), and Spear et al. (2001). Therefore, interannual variability in occurrence patterns will not be examined in detail here. Rather, data will be merged from all years of study.

Data Collection

Storm-petrel surveys.—We conducted seabird surveys during 23 cruises in the eastern Pacific Ocean, from 1980 to 1995 (Fig. 2). We counted birds from the flying bridge using the strip-transect vector method (reviewed in Spear et al. 2004). Strip widths varied, being from 400 to 600 m on one side of the ship, according to the observer's height above the water, which ranged from 12 to 16 m. Height varied, depending on the ship involved. Transect width was calibrated and maintained using the method of Heinemann (1981). During each cruise, except two, at least two observers were on watch at all times. Exceptions were autumn 1983 and 1984; data from these cruises were excluded from the density analyses because of undercounting, especially of smaller seabirds, such as storm-petrels, which occurs when surveys are conducted without multiple observers on watch simultaneously (Spear et al. 2004). We conducted continuous transects from dawn to dusk while the ship was underway. Each transect lasted 0.5 h, except while we were in waters <2,000 m deep, when transect duration was 0.25 h. By noting ship speed (km h^{-1}) and survey width, we calculated the surface area of ocean surveyed during each transect. Within the entire study area, we made 9,308.1 h of observation and surveyed 111,029 km^2 of ocean, 61,131 km^2 in boreal spring–austral autumn and 49,898 km^2 in boreal autumn–austral spring. Surveys included 768.3 h within 1,000 km of the South American coast, when >11,203.7 km^2 of ocean was surveyed, 7,382.1 km^2 in austral autumn and 3,821.6 km^2 in austral spring. Surveys within 500 km of North America were conducted over 2,557.2 h, including 12,473 km^2 of area in boreal spring and 3,061.3 km^2 in boreal autumn.

We recorded each storm-petrel, by species, and noted behavior: sitting on the water, foraging (including feeding, pattering, or circling to inspect a potential food object), or flying in a steady direction. For birds exhibiting the latter behavior, we recorded flight direction to the nearest 10°. We rarely recorded storm-petrels as having been attracted to our survey vessel; of the 36,005 petrels recorded (uncorrected count; see below), only 4 (0.01%), including two Leach's and two Wilson's storm-petrels, were recorded as having been attracted (see also Warham 1990). Observed numbers were adjusted for bird "flux" (i.e., the effect of flight speed and direction of the bird in relation to ship speed and direction) (Spear et al. 1992; flight speeds from Spear and Ainley 1997a). Adjustments yielded non-integer values. The possibility of counting the same birds twice was small, because we nearly always traveled straight through an area (i.e., we seldom made random course changes) at speeds of 19–31 km h^{-1}, which is much greater than the usual flight speed of storm-petrels (Spear and Ainley 1997a). We calculated transect density as the adjusted number of birds divided by the ocean area (km^2) surveyed during a given transect.

We had complete survey coverage of only 8 of the 26 taxa of eastern Pacific storm-petrels (Table 1). We were missing coverage of much of the north Pacific waters occupied by light-rumped Leach's Storm-Petrels; the Gulf of California (including part of the at-sea range of Black and Least storm-petrels); the California Bight (including part of the pelagic range of Ashy Storm-Petrels); waters south of 50°S (part of the range of *Oceanites o. oceanicus*, *O. o. chilensis*, likely *Fregetta grallaria segethi*, and *Pelagodroma m. dulciae–maoriana*); and the part of the pelagic range of *Nesofregetta fuliginosa* and *Oceanodroma castro* west of 170°W.

Habitat variables.—At the start of each transect, we recorded ocean depth (m), sea-surface temperature (°C) and salinity (ppt), thermocline depth (m), thermocline "strength" (see below), and wind direction and speed (km h^{-1}). Thermocline depth and thermocline strength, indices of water column mixing, were determined every 4–6 h with expendable bathythermographs. Thermocline depth is the

TABLE 1. Range limits, range coverage, survey effort, number of survey segments (*n*), and colony locations used in GAMs, abundance analyses, and ANOVA habitat analyses of 17 storm-petrel taxa surveyed in the eastern Pacific Ocean, 1980–1995. Survey effort (number of square kilometers surveyed) is given for boreal or austral spring / autumn as appropriate for northern- or southern-hemisphere breeders.

Primary taxon	Range limits		Range coverage			
	Latitude	Longitude	(km^2)	Survey effort	*n*	Population center
			Northern-hemisphere breeders			
Leach's LR	45°N–30°S	71–180°W	NC	59,689 / 49,752	2,307 / 1,802	Aleutian Islands–southeast Alaska
Leach's DR	45°N–10°S	80–140°W	C	41,384 / 37,588	1,655 / 1,446	San Benitos
Ashy	38.5°N–36.5°N	115–127°W	NC	11,332 / 3,061	4,295 / 1,524	Farallon Island
Black	40°N–15°S	76–130°W	NC	28,303 / 19,738	1,098 / 845	San Benitos
Least	32°N–5°S	120–81°W	NC	22,389 / 18,545	928 / 719	San Benitos
			Southern-hemisphere breeders			
Wedge-rumped						
O. o. tethys	30°N–25°S	75–180°W	C	41,822 / 46,494	1,378 / 1,437	Galápagos Islands
O. o. kelsalli	32°N–30°S	71–120°W	C	13,314 / 16,346	689 / 754	Peru coast
Band-rumped	25°N–25°S	71–180°W	NC	40,689 / 45,336	1,529 / 1,757	Galápagos Islands
Markham's	18°N–30°S	71–115°W	C	18,767 / 25,041	766 / 982	Peru coast
Ringed	3°S–32°S	71–85°W	C	5,975 / 9,084	248 / 342	Peru coast
White-vented	3°S–32°S	71–85°W	C	5,975 / 9,084	248 / 342	Peru coast
Wilson's						
O. o. oceanicus	15°N–5°S	110–170°W	NC	27,397 / 37,335	1,248 / 1,258	Antarctica
O. o. chilensis	13°N–54°S	71–100°W	NC	11,285 / 15,491	512 / 644	Patagonia
White-bellied						
F. g. grallaria	4°N–20°S	110–170°W	NC	30,451 / 32,833	1,201 / 1,198	Kermadec Island
F. g. segethi	4°N–36°S	71–100°W	C	8,960 / 14,523	394 / 561	Patagonia
White-throated	15°N–15°S	71–180°W	C	41,705 / 43,925	1,756 / 1,631	Christmas Island
White-faced	10°N–45°S	71–150°W	NC	30,005 / 35,373	1,086 / 1,315	New Zealand

Abbreviations: LR = light-rumped, DR = dark-rumped; NC = not complete, C = complete.

depth where the warm surface layer meets the cooler layer below (i.e., the shallowest inflection point in the temperature profile). We measured thermocline strength as the difference between temperature at the top of the thermocline and that 20 m below. A region with strong upwelling, which mixes surface waters, had a shallow, weak thermocline; the reverse occurred in a region with little mixing. We consider three bathymetric boundaries: continental shelf (ocean depth <201 m), continental slope (depth 201–2,000 m), and pelagic (depth >2,000 m).

Identification of storm-petrels at sea.—Each of the 16 species (but not all races; see below) reported on here can be distinguished at-sea on the basis of color pattern, size, and flight characteristics. Those storm-petrels that are difficult to distinguish from one another are Wilson's and White-vented, Band-rumped and Leach's, Ashy and dark-rumped Leach's, and Black and Markham's. The two races of Wedge-rumped Storm-Petrel cannot be distinguished at sea. The same is true for the two Pacific races of Band-rumped, two of the three Pacific races of White-bellied, the two Pacific races of Wilson's, and the races of White-faced storm-petrels (details given above).

White-vented Storm-Petrels can be distinguished from Wilson's by their smaller size (about two-thirds that of Wilson's) and more erratic flight. White-vented Storm-Petrels also have a partially white belly (compared with the all-dark belly of Wilson's Storm-Petrels). However, this latter character is difficult to see, because both these species fly low to the water.

The Band-rumped Storm-Petrel is darker than Leach's Storm-Petrel and has a sharply demarcated, band-shaped, pure white rump patch that shows greater contrast with the dark back and tail than the latter's V-shaped, centrally bifurcated, often smudgy white rump patch. The flight of Band-rumped Storm-Petrels tends to be swifter and more direct than that of Leach's Storm-Petrels; the former also usually fly with their wings angled back toward the tail (a flight profile similar to that of the Black), more so than the latter.

The Ashy Storm-Petrel is smaller than the dark-rumped Leach's Storm-Petrel, has more rounded wings and a more erratic, butterfly-like flight, and generally flies lower over the water than the latter.

Black and Markham's storm-petrels are large and all dark, though the former has a distinctive profile when in flight and a more direct, faster, and stronger flight than the latter and usually flies <1 m above the ocean surface. By contrast, Markham's Storm-Petrels often rise to heights >1 m and have a more leisurely flight style, similar to that of Leach's Storm-Petrels.

Assessment of the annual cycle.—Storm-petrels were collected during spring and autumn each year, 1983–1991. Besides diet, the results for which will be presented elsewhere (Spear et al. 2007), we used attributes of specimens to gauge the progression of the annual cycle (e.g., gonad development, molt, etc.). In presenting our results, we do not mention annual cycle unless we obtained data that changes what is contained in the literature reviewed above. The speed of molt and its overlap with breeding have been shown to differ between sedentary and migratory storm-petrels (Ainley et al. 1976).

At ocean stations where an inflatable boat (5 m long, with 20–35 hp motor) could be deployed, bird sampling was conducted using a shotgun. Sampling locations included recovery–deployment sites of NOAA (National Oceanic and Atmospheric Administration) buoys and deep CTD (conductivity–temperature–depth) stations throughout the study area. At a given deployment station, an attempt was made to collect five birds of each storm-petrel species present in the area. This was done by driving the inflatable boat 2–3 km from the ship. The motor was stopped, and a slick was created by pouring cod-liver oil on the water. The slick was freshened periodically by the addition of oil, about every 1–2 h, depending on wind speed (and our drift), which was the primary factor causing the oil slick to break up and disperse. The fish slick was very effective in attracting storm-petrels.

All collected birds were immediately placed in a cooler that contained 10–15 kg of ice held within plastic bags. Towels covered the ice bags to keep birds dry, to facilitate accurate determination of body mass once we returned to ship (see below).

Back at the ship, birds were weighed (to the nearest gram) and measured. We did not weigh birds that had become wet below the contour (outer) feathers (i.e., with significant water retention) until they had dried. Mean values reported here are the average mass of each species after the mass of the food load is subtracted. Measurements included total length, wing extent, unflattened wing chord, tarsus, culmen and tail lengths, culmen depth at base of nare, aspect ratio, and wing load (for the latter, see Spear and Ainley 1997b).

L.B.S. also examined most individuals to determine sex, breeding status, molt condition, and fat load. Sex and breeding status were determined by examining gonads. Females were classed as having bred previously (i.e., as adults) if their oviducts were convoluted; if their oviducts were uniform in width, they were classified as not having bred (subadults). Teste width of males that have not bred previously is smaller than that of males that have bred, because testes do not recede to the original width once an individual has bred (when the testes expand several orders of magnitude; e.g., Broughton 1994). The difference between adult and subadult teste width is ≥2 mm among storm-petrels. Fledglings can be identified during the postbreeding period by their fresh plumage and complete absence of molt compared with older birds that exhibit considerable molt.

Primary molt and body molt were scored for most individuals as follows: 0 = no primary–body molt, 1 = "light" body molt or replacing primaries 1 to 3; 2 = "moderate" body molt or replacing primaries 4 to 7; 3 = heavy body molt or replacing primaries 8 to 10.

The amount of fat covering the pectoral muscles, abdomen, and legs was examined, and fat load was scored as 0 = no fat, 1 = light fat, 2 = moderate fat, 3 = moderately heavy fat, and 4 = very heavy fat (for validation of this method, see Spear and Ainley 1998).

Data Analysis

Abundance estimation.—Following the methods developed by Clarke et al. (2003), we used S-PLUS (MathSoft, Seattle, Washington) to examine the seasonal distributions and abundance of each storm-petrel species or race with generalized additive models (GAMs; Hastie and Tibshirani 1990). We used GAMs to deal with our nonrandom survey effort (including the undersurveyed areas) in combination with the nonrandom distributions of storm-petrels. Being model-based rather than sample-based, GAMs are designed to overcome such biases. They also capture complex nonlinear trends in density while using only a few parameters. Specifically, only four independent variables were initially included in each model (latitude, longitude, ocean depth, and distance to mainland). Therefore, besides increasing accuracy (with small degrees of freedom) compared with analytical methods used previously, GAMs considerably improved the precision of abundance estimates among marine biota usually characterized as having extremely variable (clumped) densities over their pelagic ranges (reviewed in Hunt 1990).

Modeling spatial distributions—A "range area" was determined for each storm-petrel species or race on the basis of geographic limits of the pelagic ranges observed during the present study (Table 1). We then fitted GAMs for each species or race using the observed storm-petrel counts during each survey segment (see below) as the response variable. Segments outside the range area of a given species or race were excluded. On the basis of segment position, ocean depth and distance to mainland were calculated using coastline and bathymetry data obtained from the National Geophysical Data Center and WORLDBATH, respectively (see Acknowledgments). In many cases, count data are modeled using a Poisson error structure, in which the variance is equal to the mean (McCullagh and Nelder 1989). However, when birds occur in clusters, as is the case with storm-petrels, the variance of the counts is more dispersed than implied by a Poisson distribution. Therefore, we modeled these data using the Poisson variance function, estimating a dispersion parameter, which we incorporated into the model-selection procedures (e.g., Venables and Ripley 1997). Observed counts had to be adjusted for bird movement and depended on the area surveyed within each segment, so we used the logarithm of the area surveyed, multiplied by the bird-movement adjustment factor (which varies for each data point) as an offset. The logarithm was used because we used a log link function.

Once fitted, a GAM provides a smooth average density surface over the area of interest, including unsampled areas. Abundance was estimated by integrating numerically under this surface. This was done by first creating a fine grid across the map of the study area. The fitted surface was then used to predict the average number of birds in each grid-square. Finally, abundance was estimated as the sum of the predicted numbers over all grid-squares within the study area.

Bootstrap variance estimation.—To control for the correlation between counts from survey segments that were close in space and time, confidence intervals for population size were obtained using an adaptation of a moving-blocks bootstrap (Efron and Tibshirani 1993), in which the data are resampled with replacement from all possible contiguous blocks of some specified length. The block lengths were chosen by taking into account the strength of the autocorrelation between survey transects, considering that block must be long enough that observations farther than one block length apart are independent.

The block length used was defined as one survey segment. The "length" of each segment was measured from the number of 0.25- to 0.5-h transects surveyed within it. The resampling algorithm worked through the data set, recreating each segment's data in turn. Generating data for a segment involved randomly selecting a segment from the survey data and randomly selecting a transect to start from within that segment. Counts for the survey transects in the original segment were then recreated, in turn, from the survey transects in the new segment using the semiparametric bootstrap procedure (e.g., Davison and Hinkley 1997) described below. If the end of a segment was reached before enough transects had been resampled, the resampling was continued at the start of the next segment. Considering all data collected in the survey area during boreal spring–austral autumn and boreal autumn–austral spring, the result was 2,307 and 1,869 survey segments, respectively. The average surface area of ocean surveyed per segment (mean ± SD) was 26.5 ± 9.1 km^2 in the former season and 26.7 ± 7.8 km^2 in the latter.

A total of 199 bootstrap resamples was generated for each data set modeled. The model was refitted to each bootstrap resample, and a new abundance estimate was obtained. The coefficient of variation (CV) of the abundance estimate was calculated by dividing the sample standard deviation of the scaled bootstrap estimates by the original abundance estimate. The 95% confidence intervals (CIs) were estimated using the percentile method (Davison and Hinkley 1997).

Habitat use.—Generalized additive models have not yet been sufficiently developed to examine interactions between two independent environmental–temporal variables. Therefore, we used multiple linear regressions in STATA (Stata Corp., College Station, Texas) to investigate storm-petrel species' relationships with oceanographic habitat variables to which birds might be responding. Independent variables were sea-surface temperature and salinity, thermocline depth and strength, and wind speed. For species whose colony location was known or that were known at least to breed on the South American coast, we controlled for the effects of colony distance and distance to land, respectively. We expected seasonal differences from the effects of habitat parameters, but not annual ones (Ribic and Ainley 1988). Therefore, we combined data over years and seasons for each species and tested interactions between the variable "season" and each of the oceanographic habitat variables listed above. Seasons (spring and autumn) reflect the boreal time-frame for taxa that breed in the northern hemisphere, and the austral time-frame for taxa that breed in the southern hemisphere. To avoid autocorrelation, the sample unit was one survey segment (details provided above). Transects were excluded from regression analyses if there were missing data for independent variables. Further details on our use of multiple regressions are contained in Spear et al. (2003).

Independent variable values used in these regression analyses were the midpoint values from each transect. For multiple-regression analyses, all habitat variables were initially entered into the model. Insignificant terms were dropped, one at a time, in order of increasing probaiblity (P) values. Because many variables were correlated (Table 2), the importance of some variables was likely masked by others in the initial model. Therefore, we tested for effects of eliminated terms by putting them, one at a time, back into the model. The model was complete if no terms could be added or dropped. In a multiple-regression model, any independent (habitat) variables that test as having a significant relationship with the dependent variable (bird density) are considered to be true influences. That is, their effects are independent of those of other habitat variables that also have significant relationships with density, because each habitat term included in the model is evaluated while taking into account (controlling for) the effects of each of the other habitat variables. Frequently, habitat variables that are correlated can each have a significant, but independent, relationship with the dependent variable. The

TABLE 2. Pearson linear correlation coefficients (r) among oceanographic variables for the Humboldt Current (n = 397 survey transects) and for the eastern tropical Pacific Ocean, excluding the Humboldt Current (n = 3,415 surveys). Values in bold are significantly correlated ($P < 0.05$).

	Season	Sea-surface temperature	Sea-surface salinity	Thermocline depth	Thermocline strength
Humboldt Current					
Sea-surface temperature	**−0.385**				
Sea-surface salinity	**−0.241**	**0.531**			
Thermocline depth	**−0.213**	**0.6**	**0.207**		
Thermocline strength	−0.013	0.027	0.046	−0.073	
Wind speed	**−0.149**	**−0.261**	0.001	−0.007	**−0.240**
Eastern tropical Pacific without Humboldt Current					
Sea-surface temperature	**−0.196**				
Sea-surface salinity	**−0.14**	**0.363**			
Thermocline depth	0.044	**0.186**	**0.246**		
Thermocline strength	**−0.054**	**0.423**	**−0.086**	0.028	
Wind speed	**0.081**	**−0.164**	0.006	**0.155**	**−0.057**

identification of such relationships, as was the case in the present study, is one of the benefits provided by multiple-regression analysis.

We log-transformed densities to satisfy assumptions of normality (skewness–kurtosis test for normality of residuals, $P > 0.05$). All analyses of variance (ANOVAs) were conducted using log-transformed density values (calculated as: log [density + 0.1]). Although the residuals were not normally distributed in all analyses, we consider this unimportant because least-squares regression (ANOVA) is very robust with respect to non-normality (Seber 1977, Kleinbaum et al. 1988). Although they yield the best linear unbiased estimator in the absence of normally distributed residuals, P values near 0.05 must be regarded with caution (Seber 1977). Therefore, we accepted significance when using ANOVA analyses at $P < 0.025$ instead of $P < 0.05$. For chi-square analyses, significance was accepted at $P < 0.05$.

We used Sidak multiple comparison tests, an improved version of the Bonferroni test (SAS Institute 1985), to compare storm-petrel density among current systems. We also conducted a principal component analysis (PCA) in conjunction with ANOVA to compare overall habitat use among storm-petrel taxa on a seasonal basis. Habitat variables were the same as those used in the regression habitat analyses. Sample size for PCA analyses was equal to the number of survey segments in which a given taxon was recorded, including 4,284 segments in boreal spring–austral autumn and 3,759 in boreal autumn–austral spring. Aside from providing information about seasonal differences in the responses among each petrel taxon, the PCA is a means of determining which of the five habitat variables were most important in affecting the hydrobatids as a group.

To test for significant differences in overall habitat use, by season, among the 15 taxa, we used two one-way ANOVAs. In the first, we tested for differences among means of the first principal component (PC1) scores of the data representing each species; in the second, we compared PC2 score means. We considered differences between two species to be significant if either or both of the mean PC1 or PC2 scores differed significantly between them.

Hydrobatid densities per survey transect were weighted in the regression analyses by the ocean area surveyed to control for differences in area surveyed per transect. The PCAs were weighted by density per transect segment. Densities are reported here as birds per 100 km^2 and, unless noted otherwise, were calculated as the adjusted number of birds divided by the area (km^2) surveyed, multiplied by 100. Unless otherwise noted, results are reported as means ± SE. In cases where species are represented by more than one race, hereafter we use scientific names.

Pelagic behavior.—Storm-petrels feed primarily at night on prey that migrate to the ocean surface (Spear et al. 2007), and form rafting groups of satiated birds during the day, especially in the morning hours (Murphy 1924; L. B.

Spear and D. G. Ainley pers. obs). They also feed actively during the day on noncephalopod invertebrates (Spear et al. 2007). Therefore, we assumed that birds recorded sitting on the water and feeding–foraging (see above) during the day, as opposed to flying directionally (i.e., in transit), were birds that that had been feeding in the general area within the past 12 h. Working from this assumption, we estimated the tendency of storm-petrels to feed in different seasons and current systems by using chi-square tests to compare the proportion of birds recorded feeding–resting as opposed to flying in transit.

Results

Hereafter, seasons (spring and autumn) reflect the boreal time-frame for taxa that breed in the northern hemisphere, and the austral time-frame for taxa that breed in the southern hemisphere. For brevity, in the accompanying figures, we identify taxa by scientific names.

Northern-hemisphere Breeders

Leach's Storm-Petrel (Light-rumped Form)

Range area and distribution.—We had incomplete coverage of the pelagic range of light-rumped Leach's Storm-Petrels, because we lacked surveys of most of the Pacific north and west of 45°N, 130°W. We recorded 11,965 of these birds, and this number translated to 11,100.1 after adjustment for bird movement in relation to the ship. We did not observe these storm-petrels south of 27.3°S; we observed them throughout the study area, but more often toward the north during both spring and autumn. Hence, we describe the distribution, abundance, and habitat use of light-rumped Leach's Storm-Petrels for waters extending from 45°N to 30°S and from the coast of the Americas to 170°W (Table 1). Assuming that individuals from breeding colonies distributed across the width of the northern Pacific fly south to reach wintering grounds, without moving along corridors along the Pacific edges, we believe that this area includes the entirety of the range for *O. l. socorroensis* and *O. l. cheimomnestes*, and perhaps a great deal of the winter range for *O.l. leucorhoa* breeding along the Pacific Coast of North America south of the Aleutian Islands (i.e., from southeast Alaska south). Supporting this assumption, no evidence exists to indicate a major passage of these birds through the California Current during spring or fall (e.g., Briggs et al. 1987).

Like Crossin (1974) and Vilchis et al. (2006), we recorded highest densities of Leach's Storm-Petrels in the easternmost Pacific, though abundance was higher in more northern waters in spring and shifted south in autumn (Fig. 3). Significantly higher abundance of light-rumped Leach's Storm-Petrels was recorded in pelagic waters than in continental shelf and slope waters (Sidak test, both $P < 0.001$; Fig. 4A). Densities did not differ significantly between slope and shelf waters ($P = 0.1$). Light-rumped Leach's Storm-Petrels occurred much farther west than the dark-rumped forms (see below).

During spring, densities of light-rumped Leach's Storm-Petrels were higher in the ECC and NEC than in the SEC and Humboldt Current (Sidak tests, all $P < 0.01$; Fig. 5A). Spring densities did not differ between the ECC and NEC ($P > 0.4$) but were greater in the SEC than in the Humboldt Current ($P < 0.001$). In autumn, densities did not differ significantly among the SEC, ECC, and NEC (Sidak tests, all $P > 0.02$), but densities in those three current systems were all greater than that of the Humboldt Current (all $P < 0.0001$).

Habitat use.—The multiple-regression model for light-rumped Leach's Storm-Petrels explained 18% of the variation in their at-sea distribution. When controlling for seasonal differences, all five environmental variables were included among the significant main effects (Table 3; similar to Vilchis et al. 2006). Distance to colony was not tested, because of the presence of a great many Leach's Storm-Petrel colonies in the northern portion of the study area. The relationship of density to sea-surface salinity, thermocline depth, and wind speed differed significantly between seasons (Table 4, and see below). Densities of light-rumped Leach's Storm-Petrels increased with increase in sea-surface temperature and thermocline depth and strength, but decreased with increase in sea-surface salinity. The quadratic relationship between density and sea-surface temperature reflects a density decrease associated with temperatures of about 14.0–27°C, followed by a gradual density decrease at higher temperatures. The quadratic effect of sea-surface salinity was attributable to an increase in density with a

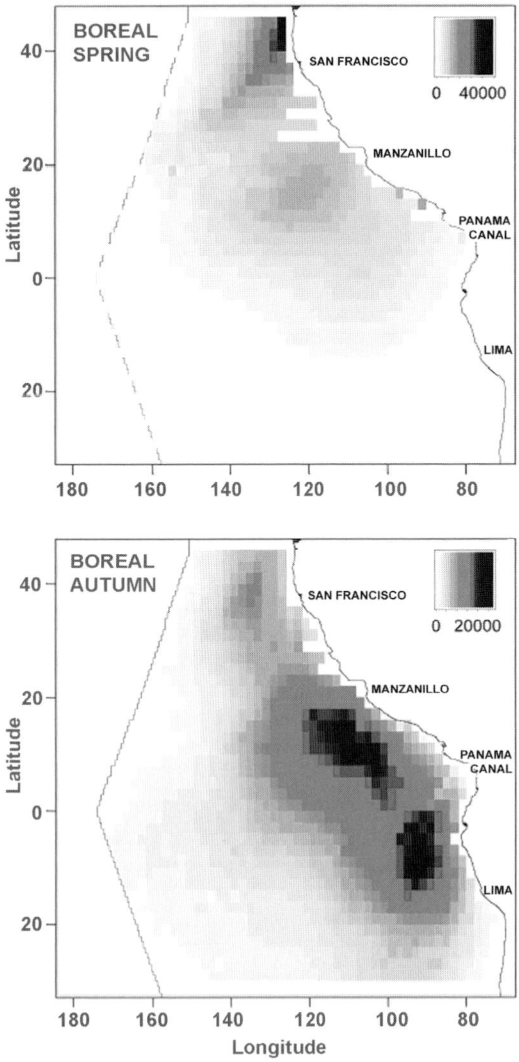

FIG. 3. Predicted distribution (birds per 2° latitude × 2° longitude cell) of light-rumped Leach's Storm-Petrels during boreal spring and autumn, 1980–1995. Numbers of birds per cell are smoothed values predicted and plotted using a GAM. Abundance estimates are given in Table 6.

salinity increase from 31 to 34.5 ppt, followed by a leveling of density at higher salinities. The quadratic effects of thermocline depth and strength reflect density increase from 0 to 100 m and from 0° temperature change to 8°C of change per 20 m, respectively, followed by density leveling at higher depth or strength. For wind speed, a quadratic effect was attributable to density increase from wind speeds of 0–25 km h^{-1}, followed by density decrease at higher wind speeds.

The significant interaction between effects of sea-surface salinity and thermocline depth in the relationship with density of light-rumped Leach's Storm-Petrels was attributable to the lack of an effect of the two variables during spring, compared with a significant decrease in density with increase in salinity and significant increase in density with increase in thermocline depth during autumn (Table 4). The interaction between season and wind speed was attributable to a stronger negative relationship between wind speed and density in spring than in autumn.

Pelagic behavior.—Overall, a greater proportion of light-rumped Leach's Storm-Petrels was observed resting–foraging (vs. flying directionally) during spring than in autumn (χ^2 = 167.11, df = 1, $P < 0.001$; Fig. 6A). During spring, a significantly lower proportion (χ^2 = 161.50, df = 2, $P < 0.001$) was observed foraging–resting in the ECC than in the SEC or NEC (Fig. 6B). During autumn, significant differences also existed among current systems for proportion foraging–resting vs. flying directionally (χ^2 = 10.77, df = 2, $P < 0.01$; Fig. 6C), though differences were not as marked as during spring.

Population estimates.—Selected models for the distribution and abundance of light-rumped Leach's Storm-Petrels during boreal spring and autumn usually included almost all the variables, though distance to the mainland was not significant during spring (Table 5). All significant independent variables were modeled as "smoothes."

Coefficients of variation for the spring and autumn abundance estimates of light-rumped Leach's Storm-Petrels were 10 and 7, respectively, which indicates that the GAMs were very successful in fitting the survey data (Table 6). Our abundance estimate of light-rumped Leach's Storm-Petrels during spring and autumn was 4,058,600 and 6,122,000, respectively (95% CI: 3.4–5.0 million in spring and 5.7–7.1 million in autumn). Thus, abundance of light-rumped Leach's Storm-Petrels was ~1.5× greater during autumn than in spring.

The difference in seasonal estimates probably resulted from migration of adults to northern Pacific breeding locations in spring, as indicated by the significantly lower proportion of adult birds as compared with subadults during spring than autumn (χ^2 = 46.12, df = 1, $P < 0.001$; Table 7).

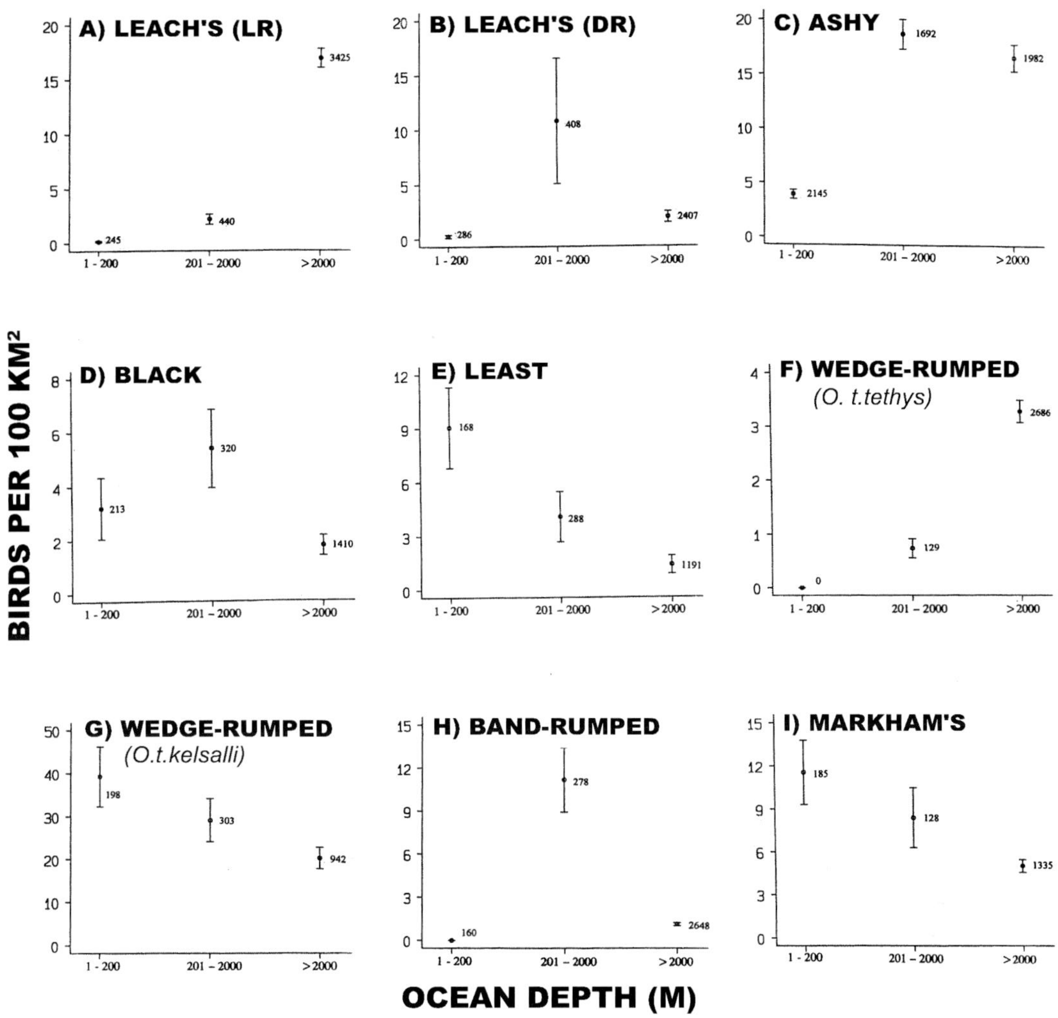

FIG. 4. Densities (birds per 100 km² of ocean surface; means ± SE) of nine taxa of storm-petrels related to ocean depth within the pelagic range of each taxon, 1980–1995. Depth 1–200 m = continental shelf; 201–200 m = continental slope; >2,000 m = pelagic waters. Numbers adjacent to means indicate number of survey segments.

Proportions of adults to subadults did not vary significantly between SEC and ECC systems in either season (chi-square tests, df = 1, $P < 0.2$). With 43% and 76% of these petrels consisting of adults during spring and autumn, respectively, estimates of their numbers in the study area during the two periods are 2.4 and 7.2 million (Table 6). Assuming that adults present in the equatorial Pacific Ocean during the breeding season were not breeding in a given year, the difference in numbers estimated during spring and autumn (4.8 million birds) is likely to be a close approximation of the breeding population. Thus, the component represented by nonbreeding adults and subadults would be ~4.7 million birds, or 49% of the population.

Leach's Storm-Petrel (Dark-rumped Form)

Pelagic distribution.—We recorded 1,205 dark-rumped Leach's Storm-Petrels in what we considered complete coverage of these forms' pelagic range; after adjustment for bird flux, the number seen was 1,127.4. We observed these birds between 44.22°N to 9.40°S. All were within 1,600 km of the Americas, except one bird

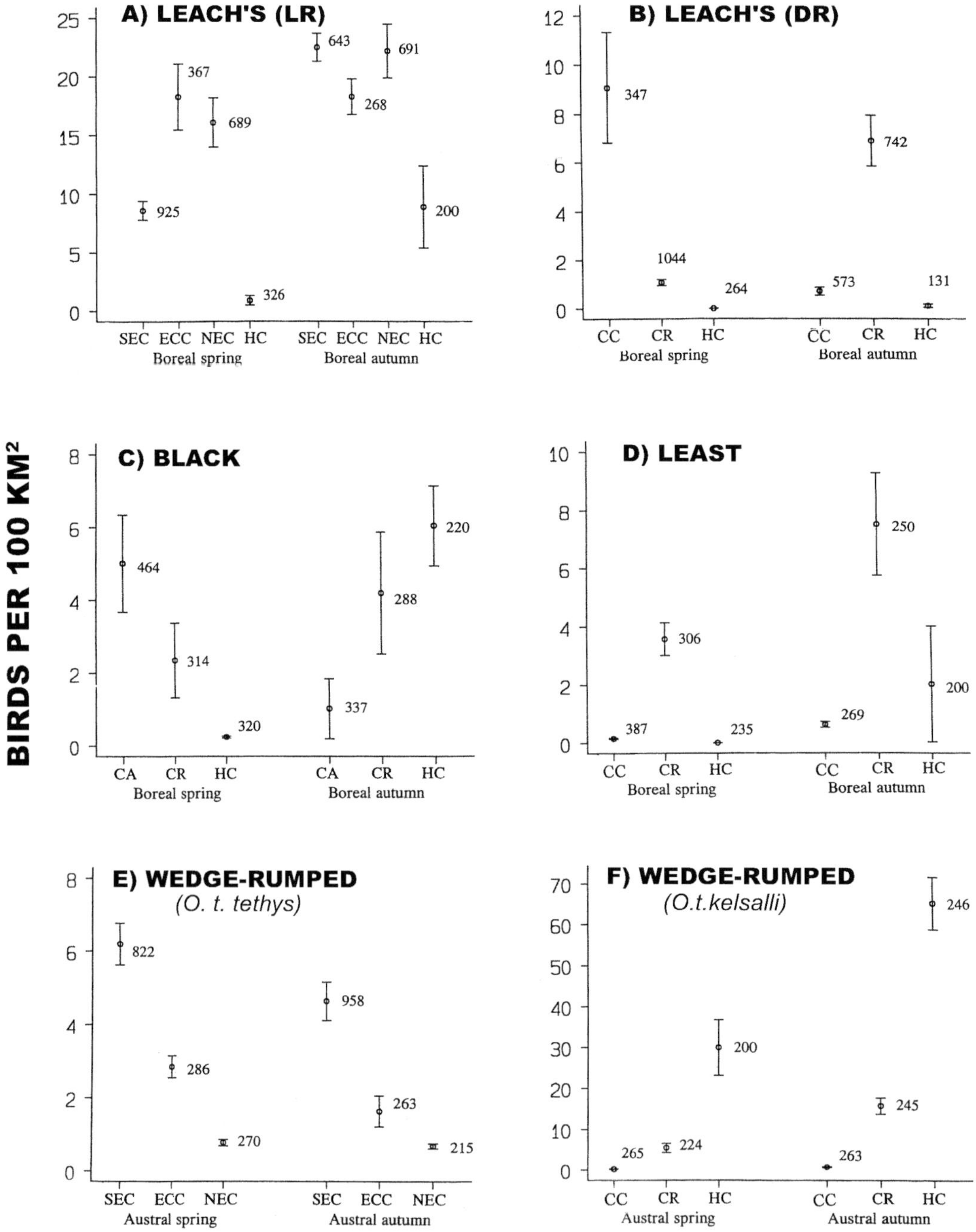

FIG. 5. Densities (birds per 100 km^2 of ocean surface; means ± SE) of six taxa of storm-petrels among the current systems occurring within the pelagic range of each taxon, 1980–1995 (SEC = Southern Equatorial Current, ECC = Equatorial Countercurrent, NEC = Northern Equatorial Current, CA = California Current, CR = Costa Rica Current, and HC = Humboldt Current). Numbers adjacent to means indicate number of survey segments.

TABLE 3. Results of multiple-regression models assessing the distribution of 17 storm-petrel taxa in the eastern Pacific Ocean in 1980–1995 in relation to significant environmental variables, the dependent variable being bird density (log birds 100 km^{-2}). Values for main effects were calculated before testing for interactions (see Table 6 for interactions); those for linear terms were calculated after polynomials for the term were removed from the model. Given for each variable are the coefficient sign (i.e., negative or positive slope, in parentheses) followed by the *P* value (x indicates <0.25, xx <0.01, and xxx <0.001); significant interactions with season are indicated with an asterisk. All numerator df = 1. Sample sizes are given in Table 1.

	Northern-hemisphere breeders				
	Leach's LR	Leach's DR	Ashy	Black	Least
VEM	18%	10%	11%	10%	9%
Sea-surface temperature					
linear	(+) xxx	(+) xxx*	(+) xxx*	(+) xxx	(+) xxx*
quadratic	(–) xxx	(–) xxx	(–) xxx	(–) xxx	NS
Sea-surface salinity					
linear	NS*	(–) xxx	(+) xxx*	(–) xxx	NS
quadratic	(–) xxx	(–) xxx	(–) xxx	(–) xxx	NS
Thermocline depth					
linear	(+) xx*	NS	NS	NS	NS
quadratic	(–) xxx	–	NS	NS	NS
Thermocline slope					
linear	(+) xxx	NS	(+) xxx	(–) xx	NS
quadratic	NS	xxx	NS	NS	NS
Wind speed					
linear	(–) xxx*	NS	(–) xx	NS	NS
quadratic	(+) xxx	NS	(–) xxx	NS	NS
Colony distance					
linear	–	(+) xxx	NS	NS	(–) xx
quadratic	–	(+) xxx	NS	(+) xxx	NS
Mainland distance					
linear	—	(–) xxx	NS	(–) xx*	(–) xxx
quadratic	—	(+) xxx	NS	(+) xxx	(+) xx

	Southern-hemisphere breeders					
	Band-rumped	Galapagos *O. t. tethys*	Galapagos *O. t. kelsalli*	Markham's	Ringed	White-vented
VEM	27%	17%	42%	21%	20%	57%
Sea-surface temperature						
linear	(+) xxx	(–) x*	(+) xxx*	(–) xx	NS	(–) xxx*
quadratic	(–) xxx	NS	NS	(–) xx	NS	(–) xxx
Sea-surface salinity						
linear	NS	(+) xx	(+) xxx	(+) xxx*	NS	NS*
quadratic	NS	(–) xx	(+) xxx	(+) xxx	(–) xxx	NS
Thermocline depth						
linear	(–) x	(–) xxx	(–) xxx	NS*	(–) xxx	NS
quadratic	(+) x	NS	NS	NS	NS	NS
Thermocline slope						
linear	NS	NS	(+) xxx	(–) xx	NS*	(–) x
quadratic	NS	xx	NS	NS	NS	(–) xxx
Wind speed						
linear	(–) xx	(–) x	(–) xxx	NS	NS	NS
quadratic	NS	(–) xxx	(–) xxx	(–) xxx	NS	NS
Colony distance						
linear	(–) xxx*	(–) xxx	NA	NA	NA	
quadratic	(+) xxx	NS	NA	NA	NA	
Mainland distance						
linear	(–) xx	(+) xxx	(–) xxx*	(–) xxx	(–) xxx	(–) xxx
quadratic	NS	(–) xxx	(–) xxx	NS	(–) xxx	(+) xxx

TABLE 3. Continued.

	Southern hemisphere breeders					
	Wilson's *O. o. oceanicus*	Wilson's *O. o. chilensis*	White-bellied *F. g. grallaria*	White-bellied *F. g. segethi*	White-throated	White-faced
VEM	NS	26%	2%	40%	5%	14%
Sea-surface temperature						
linear	NS	(–) xxx*	(–) xx	(–) xx*	NS	NS
quadratic	NS	(+) xxx	NS	(–) xxx	NS	NS
Sea-surface salinity						
linear	NS	(+) xxx	(+) xxx	NS	(+) xxx	(+) xxx*
quadratic	NS	NS	NS	NS	NS	(+) xxx
Thermocline depth						
linear	NS	NS	NS	(+) xx	(+) xxx	NS
quadratic	NS	NS	NS	NS	NS	(–) xxx
Thermocline slope						
linear	NS	NS	(+) xxx	NS	NS	(+) xx*
quadratic	NS	NS	NS	NS	NS	(–) xxx
Wind speed						
linear	NS	(–) xxx	NS	(–) xxx*	NS	NS
quadratic	NS	(–) xxx	NS	NS	NS	NS
Colony distance						
linear	NA	NA	NS	(–) xxx*	NS	(+) xxx*
quadratic	NA	NA	NS	(+) xxx	NS	(+) x
Mainland distance						
linear	NS	NS	NS	(+) xxx	NS	(+) xxx
quadratic	NS	NS	NS	(–) xxx	NS	(+) x

Abbreviations: VEM = variance explained by model; NS = nonsignificant; NA = not applicable; LR = light-rumped; and DR = dark-rumped.

observed in June 1988 at 0°, 140°W (5,000 km from the mainland) and another that we collected in April 1991 at 9°N, 140°W. Hence, we based our analyses for dark-rumped Leach's Storm-Petrels on surveys extending from 45°N to 10°S. The western cutoff was 600 km offshore for latitudes >20°N, to 1,600 km offshore for latitudes 0–20°N, and for waters within 95°W for latitudes 0–10°S. The distribution of these storm-petrels was highest off California and south of Baja California during spring (Fig. 7) but shifted farther south in autumn, with highest densities from southern Baja California to Ecuador. Densities of dark-rumped Leach's Storm-Petrels decreased significantly from the continental slope to pelagic waters (Sidak tests, both $P < 0.01$; Fig. 4B).

During boreal spring, densities of dark-rumped Leach's Storm-Petrels were significantly higher in the California Current than in the Costa Rica or Humboldt currents (Sidak tests, $P < 0.001$; Fig. 5B); densities did not differ between the latter two currents during spring ($P = 0.8$). In autumn, densities of dark-rumped Leach's Storm-Petrels were significantly higher in the Costa Rica Current than in the California and Humboldt currents (both $P < 0.001$).

Habitat use.—The multiple-regression model for dark-rumped Leach's Storm-Petrels explained 10% of the variation in their at-sea distribution (Table 3). When controlling for seasonal differences, densities of dark-rumped Leach's Storm-Petrels increased with increase in sea-surface temperature but decreased with increase in salinity and distance from land. The quadratic relationship between density and sea-surface temperature reflects a density increase associated with temperatures of 13.0–25°C, followed by a density decrease at higher temperatures. The quadratic effect of sea-surface salinity resulted from stable densities at salinities from 30 to 34 ppt, followed by a drop in density at higher salinities. The quadratic effect of thermocline strength reflects stable densities at moderate thermocline strength followed by density increase as intensity increased. The quadratic relationship of density with distance from land resulted from density decrease as distance increased to 1,000 km, followed by a leveling of density at greater distances from shore.

The significant interaction between the effects of sea-surface temperature and season reflects a strong positive relationship between density

TABLE 4. Significant interactions (indicated by an asterisk) between environmental variables and season in their relationship with storm-petrel density (log-birds 100 km^{-2}), 1980–1995. Interactions were calculated if included in Table 3. Given for the seasonal effect of each variable is the coefficient sign (i.e., negative, positive, or nonsignificant [NS] relationship with storm-petrel density, in parentheses) followed by the *P* value (xx indicates <0.01, and xxx <0.001). All numerator df = 1. Values of *n* are given in Table 1. SST = sea-surface temperature, SSS = sea-surface salinity.

Light-rumped Leach's	
Season * SSS <0.01	
Spring	(NS) 0.1
Autumn	(–) xxx
Season * thermocline depth <0.001	
Spring	(NS) 0.1
Autumn	(+) xxx
Season * wind speed <0.001	
Spring	(–) xxx
Autumn	(–) xx
Dark-rumped Leach's	
Season * SST <0.001	
Spring	(+) xxx
Autumn	(+) xxx
Black	
Season * mainland distance <0.001	
Spring	(–) xxx
Autumn	NS 0.07
Least	
Season * SST <0.025	
Spring	NS 0.08
Autumn	(+) xxx
Ashy	
Season * SST <0.001	
Spring	(+) xxx
Autumn	NS 0.7
Season * SSS <0.001	
Spring	(+) xxx
Autumn	(+) xx
Season * colony distance <0.01	
Spring	(–) xxx
Autumn	NS 0.6
White-vented	
Season * SSS <0.001	
Spring	NS 0.9
Autumn	(–) xxx
Season * SST <0.001	
Spring	(–) xx
Autumn	(–) xxx
White-bellied (*F. g. grallaria*)	
Season * SSS <0.01	
Spring	(+) xxx
Autumn	NS 0.3
Season * thermocline strength < 0.001	
Spring	(+) xx
Autumn	NS 0.1
Band-rumped	
Season * mainland distance <0.001	
Spring	(–) xxx
Autumn	(–) xx
Season * colony distance <0.01	
Spring	(–) xxx
Autumn	(–) xx
Wedge-rumped (*O. t. tethys*)	
Season * SST <0.001	
Spring	(–) xxx
Autumn	(+) xxx
Wedge-rumped (*O. t. kelsalli*)	
Season * SSS < 0.001	
Spring	(+) xxx
Autumn	(+) xx
Season * mainland distance <0.001	
Spring	NS 0.5
Autumn	(–) xxx
Markham's	
Season * SSS < 0.001	
Spring	(–) xxx
Autumn	(–) x
Season * thermocline depth <0.001	
Spring	(–) xxx
Autumn	NS 0.4
Ringed	
Season * thermocline depth <0.001	
Spring	(+) xx
Autumn	NS 0.1
Band-rumped	
Season * mainland distance <0.001	
Spring	(–) xxx
Autumn	(–) xx
Season * colony distance <0.01	
Spring	(–) xxx
Autumn	(–) <0.02
White-faced	
Season * SSS < 0.001	
Spring	NS 0.7
Autumn	(+) xx
Season * thermocline strength <0.01	
Spring	NS 0.7
Autumn	(+) xx
Season * colony distance <0.001	
Spring	NS 0.08
Autumn	(+) xx

TABLE 4. Continued.

White-bellied (*F. g. segethi*)		**Wilson's (*O. o. chilensis*)**	
Season * SST <0.01		Season * SST <0.001	
Spring	(–) xx	Spring	(–) xx
Autumn	(–) xxx	Autumn	(–) xxx
Season * wind speed <0.001		Season * wind speed <0.001	
Spring	(–) xxx	Spring	NS 0.2
Autumn	NS 0.2	Autumn	(–) xxx
Season * colony distance <0.001			
Spring	(–) xx		
Autumn	(–) xxx		

and temperature during spring compared with autumn (Table 4).

Pelagic behavior.—Overall, a greater proportion of dark-rumped Leach's Storm-Petrels was seen resting–foraging (vs. flying directionally) during boreal spring than in autumn ($\chi^2 = 49.85$, df = 1, $P < 0.001$; Fig. 6D). During spring, a significantly higher proportion of these birds was observed foraging–resting in the California Current than in the Costa Rica Current ($\chi^2 = 4.92$, df = 1, $P < 0.03$; Humboldt Current excluded, because no dark-rumped Leach's Storm-Petrels were observed there in spring; Fig. 6E). During autumn, behavior was similar when compared among the three currents ($\chi^2 = 1.02$, df = 2, $P = 0.6$; Fig. 6F).

Population estimates.—Selected models for the distribution and abundance of dark-rumped Leach's Storm-Petrels included all independent variables except longitude in both spring and autumn (Table 5). Variables were modeled as smoothes except for ocean depth (both seasons) and distance to the mainland and latitude (both modeled as linear during spring).

Coefficients of variation for the spring and autumn abundance estimates of dark-rumped Leach's Storm-Petrels were about 12 and 7, respectively, which indicates that the GAMs were successful in fitting the survey data (Table 6). Our abundance estimate of dark-rumped Leach's Storm-Petrels during spring and autumn was 682,100 and 1,200,400, respectively (95% CI: 0.6–0.9 million in spring and 1.1–1.3 million in autumn). Thus, abundance of dark-rumped Leach's Storm-Petrels was ~1.8× greater in autumn than in spring, which perhaps reflects, at least in part, the addition of recent fledglings.

Ashy Storm-Petrel

Pelagic distribution.—We had incomplete survey coverage of the pelagic range of this taxon, because of the lack of surveys off the coast of California south of Point Piños. However, we had complete coverage between 38.5°N and 36.5°N, an area in which we recorded 2,122 birds (1,880.3 birds after adjustment for bird flux). All were within 260 km of the Americas. Hence, we based our analyses for Ashy Storm-Petrels on surveys within those latitudes and ~250 km offshore (Table 1). In both seasons, the distribution of these storm-petrels was highest in waters immediately west of Southeast Farallon Island, extending north over Cordell Bank and south over Gumdrop, Pioneer, and Guide seamounts (Fig. 8). Densities were significantly higher in continental slope and pelagic waters than in shelf waters (both $P < 0.001$; Fig. 4C); densities did not differ significantly between slope and pelagic waters (Sidak test, $P > 0.1$).

Habitat use.—The multiple-regression model for Ashy Storm-Petrels explained 11% of the variation in their at-sea distribution (Table 3). Densities increased with sea-surface temperature, salinity, and thermocline strength, but decreased with increase in wind speed. The quadratic relationship between density and sea-surface salinity reflects a density increase associated with salinities of 26–34 ppt, followed by a density decrease in saltier conditions. The quadratic effect of wind speed reflects stable densities at wind speeds from 0 to 25 km h^{-1}, followed by a drop in density at higher wind speeds.

The significant interaction between the effects of sea-surface temperature and season reflects a strong positive relationship between density and temperature during spring compared with an insignificant relationship in autumn (Table 4). For salinity, the interaction with season reflects a stronger positive relationship between salinity and density in the spring compared with a weaker one in autumn. The interaction of season with distance from colony reflected

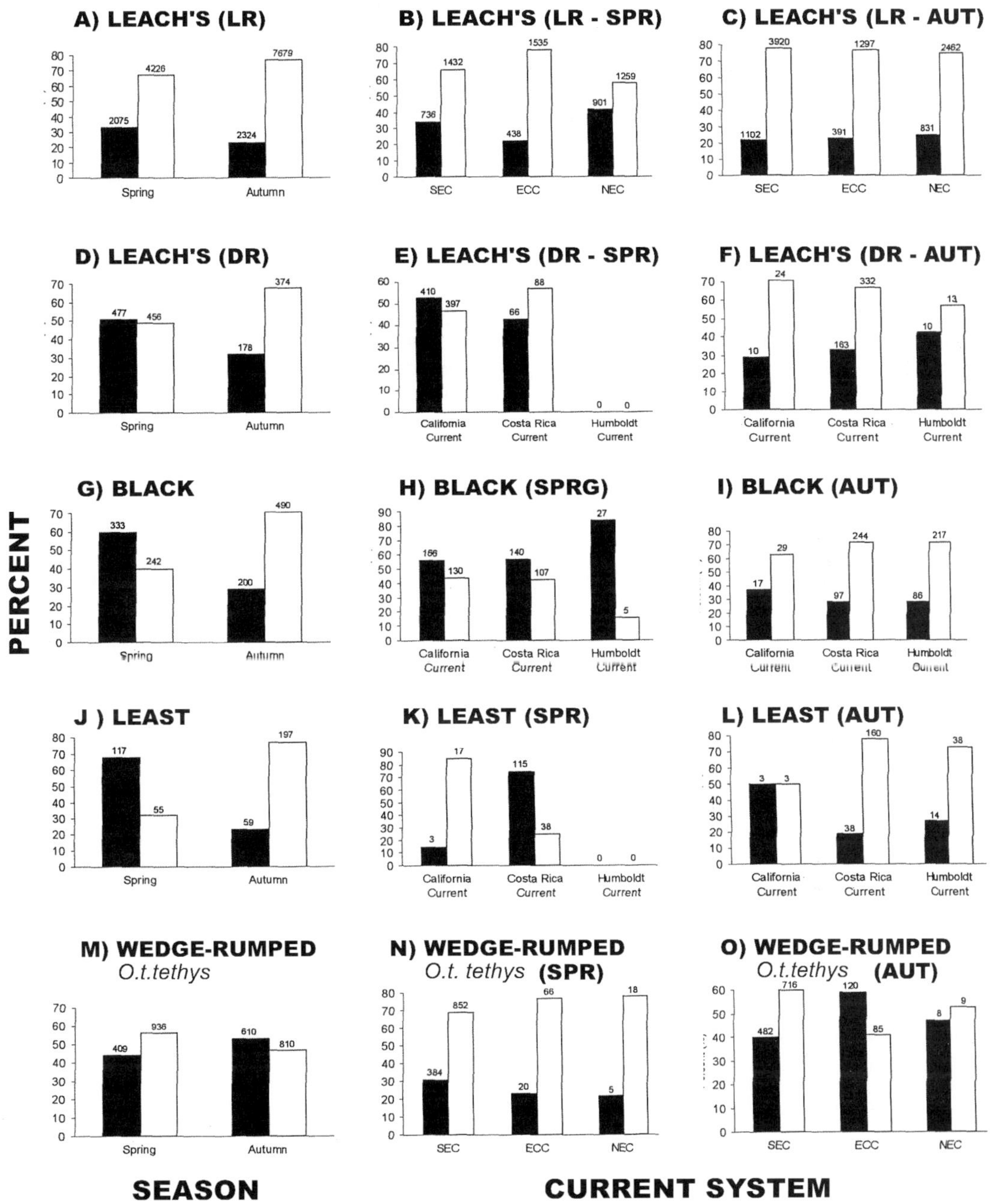

FIG. 6. Allocation of behavior among five storm-petrel taxa in different seasons and current systems, 1980–1995 (feeding–resting = light bar; in transit = dark bar). See Figure 5 for current system acronyms. Values above bars are the adjusted number of birds observed (rounded to nearest whole number).

a negative relationship of density with colony distance in spring, compared with an insignificant relationship in autumn.

Pelagic behavior.—Overall, a higher proportion of Ashy Storm-Petrels was resting–foraging (vs. flying in transit) during boreal autumn than in spring ($\chi^2 = 11.11$, df = 1, $P < 0.001$; Fig. 9).

Population estimates.—Selected models for the distribution and abundance of Ashy Storm-Petrels included all independent variables

TABLE 5. Covariates chosen by the GAMs when modeling distributions and estimating abundance of 17 storm-petrel taxa in the eastern Pacific Ocean, 1980–1995. The term "smooth" indicates that the variable was modeled as curvilinear. Seasons are boreal; NS = nonsignificant.

Season	Latitude	Longitude	Depth	Distance to mainland	Distance to colony
		Light-rumped Leach's Storm-Petrel			
Spring	Smooth	Smooth	Smooth	NS	Smooth
Autumn	Smooth	Smooth	Smooth	Smooth	Smooth
		Dark-rumped Leach's Storm-Petrel			
Spring	Linear	NS	Linear	Linear	Smooth
Autumn	Smooth	NS	Linear	Smooth	Smooth
		Ashy Storm-Petrel			
Spring	Smooth	Linear	Linear	Smooth	Smooth
Autumn	Smooth	Smooth	NS	Smooth	Smooth
		Black Storm-Petrel			
Spring	Smooth	NS	Smooth	Smooth	Linear
Autumn	Smooth	Smooth	NS	Smooth	Smooth
		Least Storm-Petrel			
Spring	Smooth	NS	Linear	Linear	Smooth
Autumn	Linear	Linear	Linear	Smooth	Smooth
		Band-rumped Storm-Petrel			
Spring	Smooth	Smooth	Smooth	Smooth	Smooth
Autumn	NS	Smooth	Smooth	NS	Linear
		Wedge-rumped Storm-Petrel (*O. t. tethys*)			
Spring	Smooth	Smooth	Smooth	Smooth	Smooth
Autumn	Smooth	Linear	Smooth	Smooth	Smooth
		Wedge-rumped Storm-Petrel (*O. t. kelsalli*)			
Spring	Smooth	Smooth	Linear	Smooth	Linear
Autumn	Smooth	Smooth	Smooth	Smooth	Smooth
		Ringed Storm-Petrel			
Spring	Linear	Smooth	NS	Smooth	Linear
Autumn	Linear	Linear	Smooth	Linear	Smooth
		Markham's Storm-Petrel			
Spring	Smooth	NS	Linear	Smooth	Smooth
Autumn	Smooth	Smooth	Smooth	Smooth	Smooth
		White-vented Storm-Petrel			
Spring	Smooth	NS	Smooth	Linear	Smooth
Autumn	Smooth	NS	NS	Smooth	Linear
		Wilson's Storm-Petrel (*O. o. oceanicus*)			
Both seasons	Smooth	Linear	NS	Smooth	Smooth
		Wilson's Storm-Petrel (*O. o. chilensis*)			
Spring	Linear	Linear	NS	NS	NS
Autumn	Smooth	NS	Linear	Smooth	Smooth
		White-bellied Storm-Petrel (*F. g. grallaria*)			
Spring	Linear	NS	Linear	NS	Linear
		White-bellied Storm-Petrel (*F. g. segethi*)			
Spring	Smooth	Smooth	Smooth	Smooth	Smooth
Autumn	Smooth	Smooth	Smooth	Smooth	Smooth
		White-throated Storm-Petrel			
Spring	Smooth	Smooth	Smooth	Smooth	Smooth
Autumn	Smooth	Linear	Linear	NS	Linear
		White-faced Storm-Petrel			
Spring	Smooth	NS	Smooth	Smooth	NS
Autumn	Smooth	NS	Linear	Linear	NS

TABLE 6. Results of generalized additive models performed to estimate abundance of storm-petrels during spring and autumn in the eastern Pacific Ocean, 1980–1995. Included are 95% confidence intervals (CI) and coefficients of variation × 100 (CV). Survey effort is given in Table 1.

Population	Size estimate	CI	CV
Northern-hemisphere residents			
Leach's Storm-Petrel LR			
Boreal spring	4,058,600	3,404,000–4,989,000	10.5
Boreal autumn	6,122,000	5,666,000–7,102,000	7.4
Leach's Storm-Petrel DR			
Boreal spring	682,100	592,700–860,500	11.6
Boreal autumn	1,200,400	1,058,200–1,333,900	6.9
Black Storm-Petrel			
Boreal spring	441,400	347,300–573,200	16.6
Boreal autumn	561,700	483,300–687,000	8.2
Ashy Storm-Petrel (Point Pinos to Bodega Bay)			
Boreal spring	7,287	4,500–9,070	18.9
Boreal autumn	4,207	2,690–6,425	24
Least Storm-Petrel			
Boreal spring	195,100	171,900–257,600	18.3
Boreal autumn	308,900	249,500–386,200	15.9
Southern-hemisphere residents			
Band-rumped Storm-Petrel			
Austral spring	122,700	95,400–140,000	11
Austral autumn	475,700	340,500–600,300	15.1
Wedge-rumped Storm-Petrel (*O. t. tethys*)			
Austral spring	1,136,900	848,700–1,449,400	12.4
Austral autumn	628,000	552,900–786,500	9.1
Wedge-rumped Storm-Petrel (*O. t. kelsalli*)			
Austral spring	1,258,700	924,300–1,478,000	8.5
Austral autumn	1,451,800	979,700–2,012,700	10.5
Markham's Storm-Petrel			
Austral spring	806,500	682,800–1,034,100	12.2
Austral autumn	1,100,000	1,028,500–1,620,300	11.9
Ringed Storm-Petrel			
Austral spring	637,200	543,500–784,000	15.5
Austral autumn	1,011,900	857,800–1,551,400	17.8
White-vented Storm-Petrel			
Austral spring	396,400	343,100–473,700	12.6
Austral autumn	858,700	811,100–1,026,300	7.5
Wilson's Storm-Petrel (*O. o. oceanicus*)			
Both seasons	8,054	6,380–12,590	22.4
Wilson's Storm-Petrel (*O. o. chilensis*)			
Austral spring	866,600	782,700–1,154,000	12
Austral autumn	406,000	295,200–562,900	17.6
White-bellied Storm-Petrel (*F. g. grallaria*)			
Austral spring	2,866	2,198–5,713	25.8
White-bellied Storm-Petrel (*F. g. segethi*)			
Austral spring	114,600	83,900–129,400	11
Austral autumn	442,500	308,900–545,400	12.1
White-throated Storm-Petrel			
Austral spring	52,300	35,510–87,900	18
Austral autumn	20,825	15,710–26,370	16
White-faced Storm-Petrel			
Austral spring	6,390	4,450–10,400	24.5
Austral autumn	713,900	569,900–884,700	13.6

TABLE 7. Summary of various within-species comparisons in relation to age in storm-petrels, as determined from gonad and oviduct condition, 1984–1995.

A. Percentage (in parentheses) of Leach's Storm-Petrels by age class occurring in the South Equatorial Current (SEC) and Equatorial Countercurrent (ECC).

	Spring			Autumn		
	SEC	ECC	Overall	SEC	ECC	Overall
Adults	39 (44)	7 (35)	46 (43)	153 (74)	56 (82)	209 (76)
Subadults	49 (56)	13 (65)	62 (57)	53 (26)	12 (18)	65 (24)
Age >1 year	49 (56)	13 (65)	62 (57)	45 (22)	10 (15)	55 (20)
Fledglings	0 (0)	0 (0)	0 (0)	8 (4)	2 (3)	10 (4)
Total collected	88	20	108	206	68	274

B. Percentage (in parentheses) of Wedge-rumped Storm-Petrels (*O. t. tethys*) that occur at various distances from Galápagos Islands by season.

	0–2000 km	2,001–4,000 km	>4,000 km	Overall
		Spring		
Adults	49 (73)	8 (62)		57 (71)
Subadults	18 (27)	5 (39)		23 (29)
Age >1 year	18 (27)	5 (39)		23 (29)
Fledglings	0 (0)	0 (0)		0 (0)
Total collected	67	13		80
		Autumn		
Adults	93 (82)	19 (65)	42 (79)	154 (79)
Subadults	20 (18)	10 (35)	11 (21)	41 (21)
Age >1 year	16 (14)	9 (31)	11 (21)	36 (20)
Fledglings	4 (4)	1 (3)	0 (0)	5 (3)
Total collected	113	29	53	195

C. Percentage (in parentheses) of various populations occurring in different parts of the species' ranges in the eastern Pacific.

	Wedge-rumped Storm-Petrel (*O. t. kelsalli*; seasons grouped)	
	Panama and Ecuador	Peru
Adults	2 (20)	15 (65)
Subadults	8 (80)	8 (35)
Total collected	10	23

	Markham's Storm-Petrel (seasons grouped)	
	<200 km from mainland	>500 km from mainland
Adults	6 (86)	1 (13)
Subadults	1 (14)	7 (87)
Total collected	7	8

	Wilson's Storm-Petrel (seasons grouped)	
	O. o. oceanicus, central Pacific	*O. o. chilensis*, Ecuadorian coast
Adults	5 (83)	2 (29)
Subadults	1 (17)	5 (71)
Total collected	6	7

	White-throated Strom-Petrel (seasons grouped)		
	110°W	140–150°W	Overall
Adults	6 (60)	8 (89)	14 (74)
Subadults	4 (40)	1 (11)	5 (26)
Total collected	10	9	19

TABLE 7. Continued.

	White-faced Storm-Petrel (austral autumn)		
	East of 110°W	130–150°W	Overall
Adults	5 (63)	7 (100)	12 (80)
Subadults	3 (27)	0 (0)	3 (20)
Total collected	8	7	15

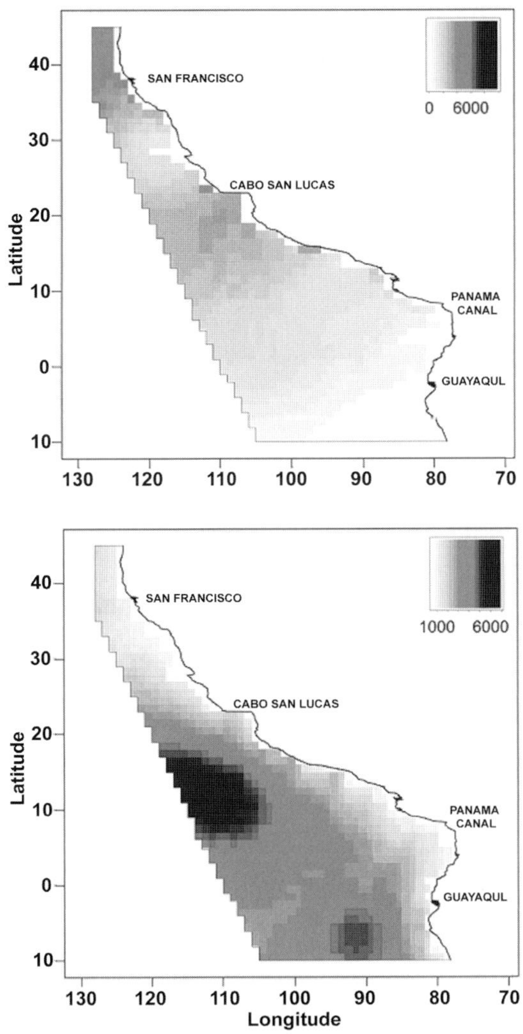

FIG. 7. Predicted distribution (birds per 1° latitude × 1° longitude cell) of dark-rumped Leach's Storm-Petrels during boreal spring and autumn, 1980–1995. See Figure 3 for further details.

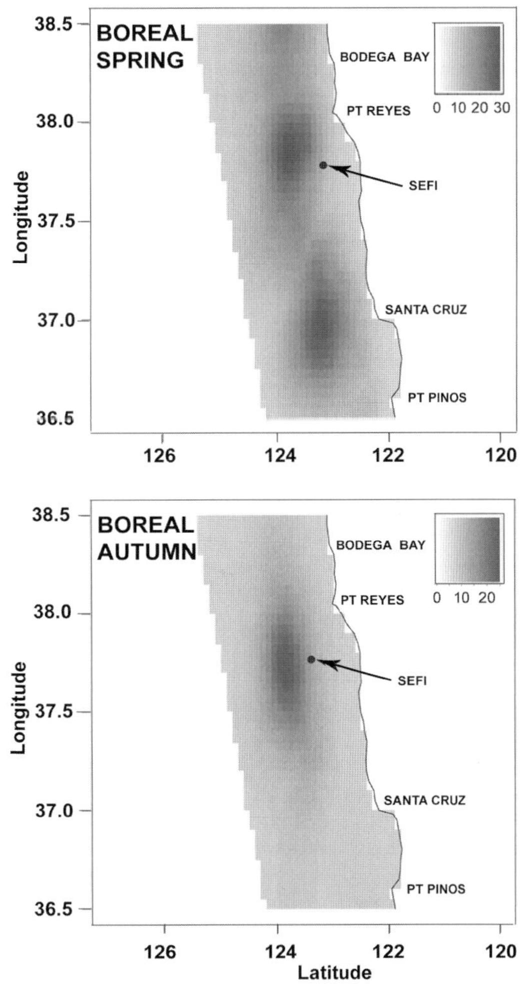

FIG. 8. Predicted distribution (birds per 0.05° latitude × 0.1° longitude cell) of Ashy Storm-Petrels during boreal spring and autumn, 1980–1995. See Figure 3 for further details.

except longitude in both spring and autumn (Table 5). Variables were modeled as smoothes except for ocean depth (both seasons) and distance to the mainland and latitude (both modeled as linear during spring).

Coefficients of variation for the spring and autumn abundance estimates were about 12 and 7, respectively, which indicates that the GAMs were successful in fitting the survey data (Table 6). Our abundance estimates of Ashy Storm-Petrels during spring and autumn were 7,287 and 4,207, respectively (95% CI: 4,500–9,100 in spring and 2,700–6,400 in autumn). Thus, estimated abundance was ~1.7× greater during spring than in autumn, which perhaps reflects a greater concentration in our study area during the peak of nesting.

Black Storm-Petrel

Pelagic distribution.—We had incomplete coverage of the pelagic range of the Black Storm-Petrel, because there are no surveys from the Gulf of California. We recorded 1,228 birds (1,136.1 after adjustment for bird flux). We observed these birds between 38.03°N and 12.65°S. All were within 360 km of the Americas. Hence, we based our analyses for Black Storm-Petrels on surveys extending from 40°N to 15°S (Table 1). The western cutoff was 400 km. During spring, abundance of Black Storm-Petrels was highest off California and Baja California (Fig. 10). In autumn, abundance was highest from Mexico and Central America to southern Peru. Significantly higher densities of Black Storm-Petrels were recorded over the continental slope compared with continental shelf or pelagic waters (Sidak test, both $P < 0.001$; Fig. 4D); densities differed insignificantly between pelagic and shelf waters ($P = 0.1$).

Spring densities of Black Storm-Petrels were similar ($P = 0.2$) in the California and Humboldt currents, both of which had higher densities than the Costa Rica Current (Sidak tests, $P < 0.025$; Fig. 5C). In autumn, density of Black Storm-Petrels did not differ among the three current systems (all $P > 0.2$).

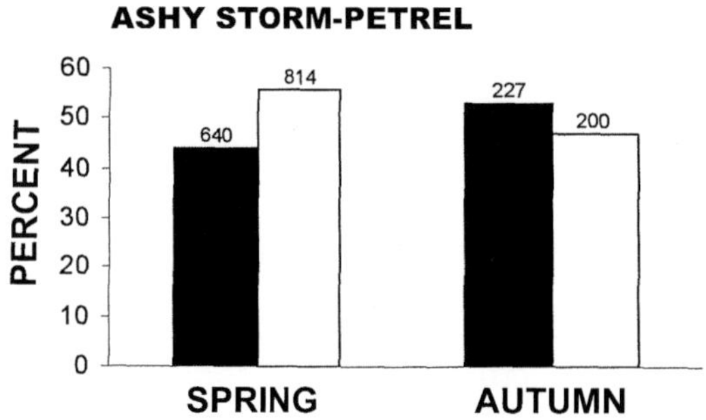

FIG. 9. Allocation of behavior among Ashy Storm-Petrels during boreal spring and autumn, 1980–1995 (feeding–resting = light bar; in transit = dark bar). Values above bars are the adjusted numbers of birds observed (rounded off to nearest whole number).

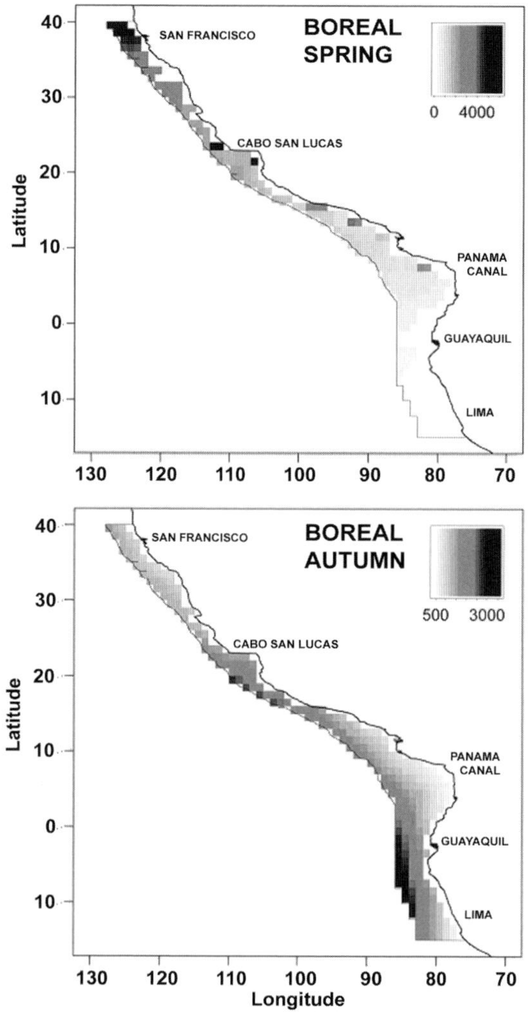

FIG. 10. Predicted distribution (birds per 1° latitude × 1° longitude cell) of Black Storm-Petrels during boreal spring and autumn, 1980–1995. See Figure 3 for further details.

Habitat use.—The multiple-regression model for Black Storm-Petrels explained 10% of the variation in their at-sea distribution (Table 3). When controlling for seasonal differences, densities of these petrels increased with sea-surface temperature but decreased with increase in salinity and thermocline strength and distance to land. The quadratic relationship between density and sea-surface temperature reflects an increase associated with temperatures of 13.0–24°C, followed by a decrease at higher temperatures (Table 4). The quadratic effect of sea-surface salinity reflects a density increase as salinities increased from 30 to 34.5 ppt, followed by a drop in density at higher salinities. The quadratic effect of colony distance results from a decline in density as distance increased to 1,500 km, followed by density increase at greater distances. The interaction between the effect of season and distance from land is attributable to a significant density decline with distance during spring, compared with a lack of relationship between density and distance during autumn (Table 4)

Pelagic behavior.—Overall, a greater proportion of Black Storm-Petrels was resting–foraging (vs. flying in transit) during boreal spring than in autumn ($\chi^2 = 107.64$, df = 1, $P < 0.001$; Fig. 6G). During spring, behavior differed significantly among current systems ($\chi^2 = 9.75$, df = 1, $P < 0.03$; Fig 6H), largely because a higher proportion of these birds was observed foraging–resting in the Humboldt Current than in the California and Costa Rica currents. However, this comparison was hampered by a low number of birds observed during spring in the Humboldt Current. During autumn, the tendency to feed was similar when compared among the three currents ($\chi^2 = 1.52$, df = 2, $P = 0.5$; Fig. 6I).

Population estimates.—Selected models for the distribution and abundance of Black Storm-Petrels included all independent variables except longitude in spring and ocean depth in autumn (Table 5). Variables were modeled as smoothes except for distance to colony, which was modeled as linear during spring.

Coefficients of variation for the spring and autumn abundance estimates of Black Storm-Petrels were about 17 and 8, respectively, which indicates that the GAMs were successful in fitting the autumn survey data but only moderately so for the spring data (Table 6). Our abundance estimates of Black Storm-Petrels during spring and autumn were 441,400 and 561,700, respectively (95% CI: 0.35–0.57 million in spring and 0.48–0.69 million in autumn). Thus, estimated abundance was 1.3× greater in autumn than in spring, which perhaps reflects the addition of fledglings or the fact that we did not survey the Gulf of California, where some number would be during spring.

Least Storm-Petrel

Pelagic distribution.—We had incomplete survey coverage of the pelagic range of this species because of the lack of surveys in the Gulf of California. Between 31.9°N and 4.9°S, we recorded 386 birds (369.0 after adjustment for bird flux). All were within 327 km of the Americas. Hence, we based our analyses for this species on surveys within those latitudes and to ~350 km offshore (Table 1). Abundance was highest from Costa Rica to the northern Panama Bight during spring; in autumn, they were most abundant from southern Baja California to Costa Rica (Fig. 11). Densities of Least Storm-Petrels decreased significantly from the continental shelf to continental slope to pelagic waters (Sidak test, both $P < 0.025$; Fig. 4E).

During both spring and autumn, densities of Least Storm-Petrels were higher in the Costa Rica Current than in the California and Humboldt currents (Sidak tests, all $P < 0.01$; Fig. 5D); densities differed little between the latter two currents in both seasons (both $P > 0.5$).

Habitat use.—The multiple-regression model for Least Storm-Petrels explained 9% of the variation in their at-sea distribution (Table 3). Densities increased with sea-surface temperature, and decreased with increase in distance from land and colony. The quadratic relationship between density and distance from land reflects a density decline with increase in distance to 200 km, followed by a leveling in density at greater distances.

The significant interaction between the effects of sea-surface temperature and season reflects a strong positive relationship between density and temperature during autumn compared with an insignificant relationship in spring (Table 4).

Pelagic behavior.—Overall, a greater proportion of Least Storm-Petrels was resting–foraging (vs. flying in transit) during spring than in autumn ($\chi^2 = 85.95$, df = 1, $P < 0.001$; Fig. 6J).

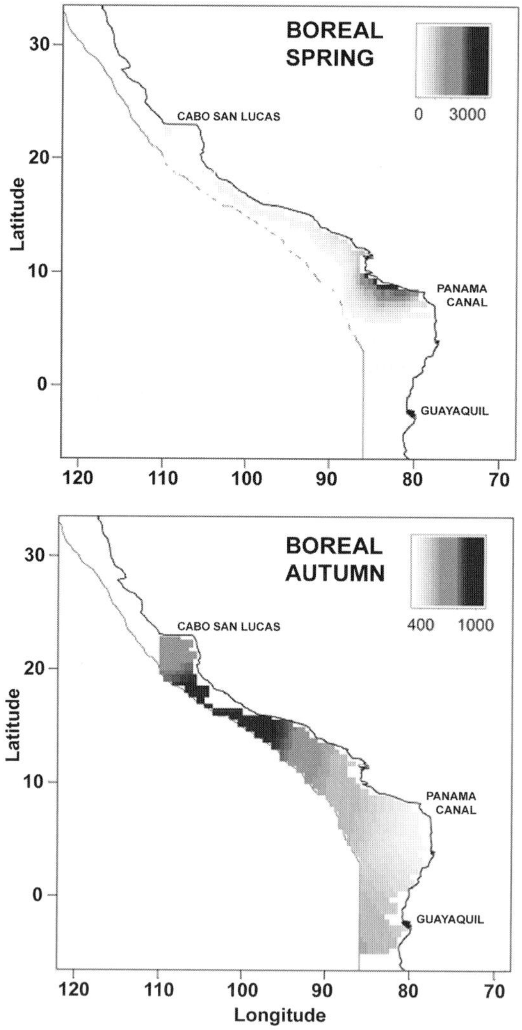

FIG. 11. Predicted distribution (birds per 0.5° latitude × 0.5° longitude cell) of Least Storm-Petrels during boreal spring and autumn, 1980–1995. See Figure 3 for further details.

During spring, a significantly higher proportion of these birds was observed foraging–resting in the Costa Rica Current than in the California Current ($\chi^2 = 29.52$, df = 2, $P < 0.001$; Humboldt Current excluded, because no Least Storm-Petrels were observed there in spring; Fig. 6K). During autumn, behavior was similar when compared among the three currents ($\chi^2 = 4.42$, df = 2, $P = 0.1$; Fig. 6L).

Population estimates.—Selected models for the distribution and abundance of Least Storm-Petrels included all independent variables except longitude in spring (Table 5). Variables were modeled as linear except for ocean depth (both seasons), latitude and longitude (both modeled as smoothes in autumn), and distance to land (modeled as smooth in spring).

Coefficients of variation for the spring and autumn abundance estimates of Least Storm-Petrels were about 18 and 16, respectively, which indicates that the GAMs were only moderately successful in fitting the survey data (Table 6). Our abundance estimates of Least Storm-Petrels during spring and autumn were 195,100 and 308,900, respectively (95% CI: 171,900–257,600 in spring and 249,500 386,200 in autumn). Thus, abundance was ~1.6× greater during autumn than in spring.

SOUTHERN-HEMISPHERE BREEDERS

Wedge-rumped Storm-Petrel (*Oceanodroma tethys tethys*)

Pelagic distribution.—We did not collect any birds of the South American form (*O. t. kelsalli*) at distances >400 km from the coast between Panama and southern Peru, nor did we collect any of the Galápagos Islands form (*O. t. tethys*) <100 km offshore or over ocean depths of <1,000 m (see Crossin [1974] for similar observations).

Judging from ratios of the two subspecies collected at different distances off the South American, Mexican, and Central American coasts (see above), there appears to be an overlap in distribution of the two forms in waters at 250–500 km offshore. Therefore, we estimated abundance and distributions of the two forms assuming a 400-km offshore cutoff between the two races (i.e., at approximately the western edge of the Humboldt Current). Using this cutoff, we saw 2,241 *O. t. tethys* (adjusted = 2,085.8) between 27.27°N and 21.18°S and west to 175.73°W. Hence, our analyses for this species pertain to waters from 30°N to 25°S and to 180°W. In both spring and autumn, *O. t. tethys* was concentrated within 2,000 km of the Galápagos Islands (Fig. 12), and the pelagic range was very similar in both seasons (see also Vilchis et al. [2006], who did not include the range of *O.t. kelsalli* in their study area). This storm-petrel was significantly more abundant over pelagic waters than over continental shelf and slope waters (Sidak tests, both $P < 0.001$;

Fig 4F); density differed little between shelf and slope habitats ($P > 0.3$).

During both spring and autumn, densities of *O. t. tethys* were highest in the SEC and declined significantly from ECC to NEC (Sidak tests, all $P < 0.001$; Fig. 5E).

Habitat use.—The multiple-regression model for *O. t. tethys* explained 17% of the variation in their at-sea distribution (Table 3). Densities of this taxon increased with sea-surface salinity and mainland distance and decreased with increase in sea-surface temperature, thermocline depth and strength, wind speed, and colony distance (see also Vilchis et al. 2006). The quadratic relationship between density and sea-surface salinity reflects a density increase to salinities of 34 ppt, followed by a leveling of density at higher salinities. The quadratic effect of thermocline strength reflects an increase in density as thermocline strength increased to a 12°C decrease in the first 20 m below the thermocline, followed by a density decline with increase in thermocline strength. The quadratic effect of wind speed reflects stable densities at wind speeds of 0–25 km h^{-1}, followed by a decline in density associated with higher wind speeds.

The significant interaction between the effects of season and sea-surface temperature reflects an increase in density with temperature during autumn, compared with a density decline with increased temperature during spring (Table 4).

Pelagic behavior.—The proportion of *O. t. tethys* seen that was feeding–resting (vs. flying in transit) was significantly greater in austral autumn than in spring ($\chi^2 = 46.74$, df = 1, $P < 0.001$; Fig. 6M). During austral spring, the proportion of birds feeding–resting (vs. flying in transit) was similar in the three equatorial current systems ($\chi^2 = 3.15$, df = 2, $P = 0.2$; Fig. 6N). However, in autumn, the proportions of birds feeding–resting (vs. flying in transit) differed significantly among the current systems ($\chi^2 = 83.53$, df = 2, $P < 0.001$; Fig. 6O), because of a greater tendency to feed when in the ECC than in the SEC and NEC.

Population estimates.—Selected models for the distribution and abundance of *O. t. tethys* included all independent variables in both seasons (Table 5); all variables were modeled as smoothes except longitude, which was modeled as linear during austral autumn.

Coefficients of variation for the austral spring and autumn abundance estimates of *O. t. tethys* were about 12 and 9, respectively, which indicates that the GAMs were very successful in fitting the survey data (Table 6). Our abundance estimates of *O. t. tethys* during spring and autumn were 1,136,900 and 628,000, respectively (95% CI: 0.8–1.4 million in spring and 0.5–0.8 million in autumn). Thus, abundance was ~1.8× greater during spring than in autumn. This difference was unexpected, given that the breeding

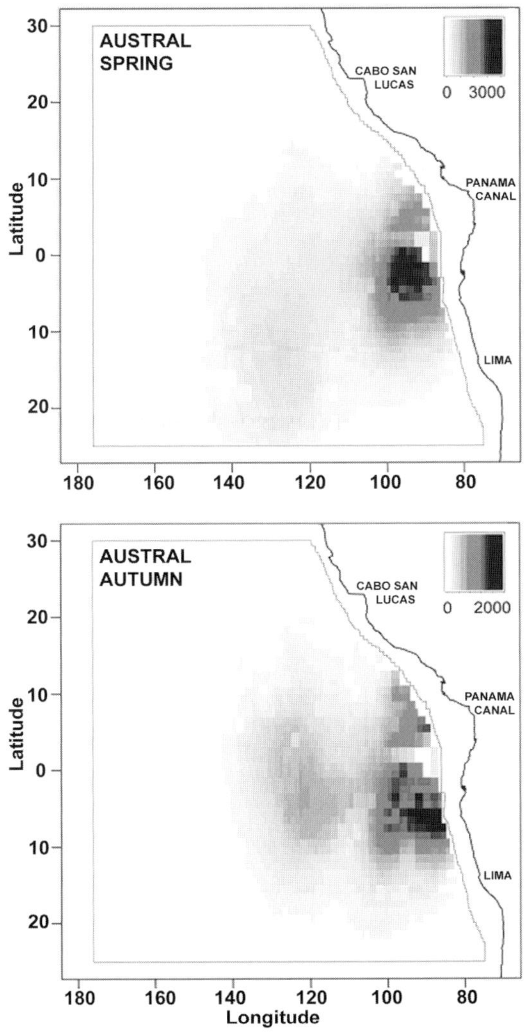

FIG. 12. Predicted distribution (birds per 1° latitude × 1° longitude cell) of the *O. t. tethys* form of Wedge-rumped Storm-Petrel during austral spring and autumn, 1980–1995. See Figure 3 for further details.

season of this taxon is believed to fall primarily within spring and summer (Harris 1969).

Ratios of adults to subadults did not vary significantly between seasons ($\chi^2 = 2.32$, df = 1, $P = 0.13$; Table 7) or with distance from the Galápagos Islands (chi-square tests for spring and autumn: $P = 0.8$ and 0.13, respectively).

Wedge-rumped Storm-Petrel (*Oceanodroma tethys kelsalli*)

Pelagic distribution.—Using the offshore cutoff discussed above with regard to *O. t. tethys*, we recorded 6,494 *O. t. kelsalli* (adjusted = 6,143.5) between 31.58°N and 28.77°S. Therefore, our analyses for this species pertain to waters from 32°N to 30°S and to 400 km offshore (Table 1). During the breeding season of *O. t. kelsalli* (austral autumn), individuals of this taxon were concentrated from southern Ecuador to central Peru (Fig. 13), with moderate numbers spread north to central Mexico. In spring, this taxon was also concentrated between southern Ecuador and southern Peru, with moderate numbers spreading into the Panama Bight. Density was significantly greater over continental shelf than over pelagic waters (Sidak test, $P < 0.001$; Fig. 4G), but densities over the slope did not differ from densities over the shelf or pelagic waters (both $P > 0.2$)

During both spring and autumn, densities of *O. t. kelsalli* were highest in the Humboldt Current and declined significantly from the Costa Rica Current to the California Current (Sidak tests, all $P < 0.001$; Fig. 5F).

Habitat use.—The multiple-regression model for *O. t. kelsalli* explained 42% of the variation in this form's at-sea distribution (Table 3). Densities of this taxon increased with sea-surface temperature and salinity and thermocline strength but decreased with increase in thermocline depth, wind speed, and distance from land. The quadratic relationship between density and sea-surface salinity results from stable density associated with salinities of 32–33.5 ppt, followed by a density increase with greater salinity. The quadratic effect of wind speed results from stable density in winds of 0–25 km h^{-1}, followed by density decrease with higher wind speeds. The quadratic effect of mainland distance results from density decrease at distances to 200 km followed by an even greater decrease at greater distances.

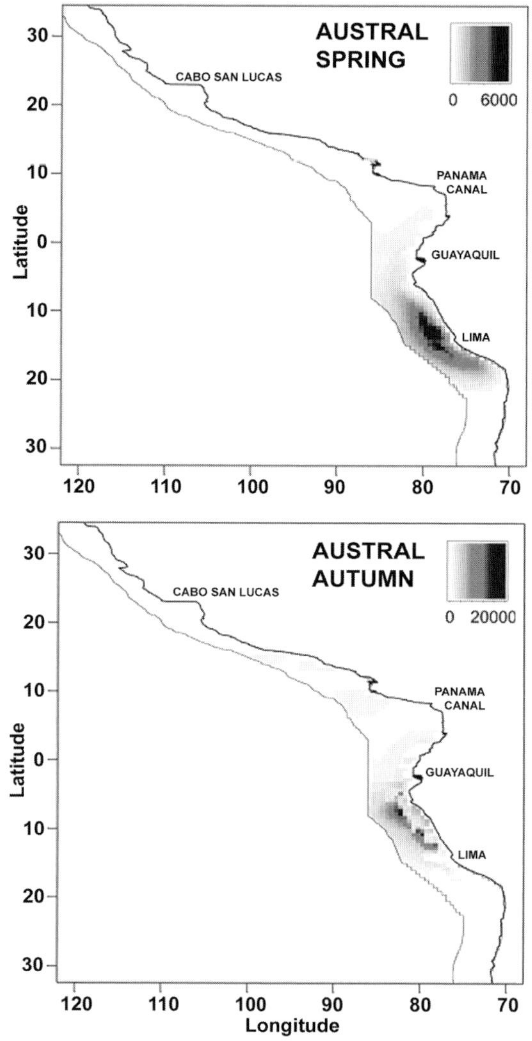

FIG. 13. Predicted distribution (birds per 0.5° latitude × 0.5° longitude cell) of the *O. t. kelsalli* form of Wedge-rumped Storm-Petrel during austral spring and autumn, 1980–1995. See Figure 3 for further details.

The significant interaction between the effect of season and sea-surface temperature reflects a stronger increase in density with temperature during austral spring than in autumn (Table 4). The interaction of season with mainland distance reflects a strong decrease in density with distance from the mainland in autumn, compared with no effect of distance on density during spring.

Pelagic behavior.—Overall, a greater proportion of *O. t. kelsalli* was resting–foraging (vs.

flying in transit) during austral autumn than in spring (χ^2 = 306.92, df = 1, P < 0.001; Fig. 14A). During austral spring, a significantly higher proportion of these birds was observed foraging–resting in the Humboldt Current than in the Costa Rica Current (χ^2 = 74.42, df = 1, P < 0.001; California Current excluded, because no *O. t. kelsalli* were observed there in spring; Fig. 14B). During autumn, behavior also differed significantly among the three boundary currents (χ^2 = 355.12, df = 2, P < 0.001; Fig. 14C), primarily because of a greater tendency for *O. t. kelsalli* to feed when in the Humboldt Current.

Population estimates.—Selected models for the distribution and abundance of *O. t. kelsalli* included all independent variables in both seasons (Table 5); all variables were modeled as smoothes except ocean depth and colony distance, which were modeled as linear during austral spring.

Coefficients of variation for the austral spring and autumn abundance estimates of *O. t. kelsallis* were about 9 and 11, respectively, which indicates that the GAMs were very successful in fitting the survey data (Table 6). Our abundance estimates of *O. t. kelsallis* during spring and autumn were 1,258,700 and 1,451,800, respectively (95% CI: 900,000–1,500,000 in spring and 1–200,000 in autumn). Thus, abundance was ~1.2× greater during austral autumn than in spring. This difference was unexpected, given that the breeding season of this taxon is believed to center in autumn (Murphy 1936).

Ratios of adults to subadults did not vary significantly between seasons (χ^2 = 2.32, df = 1, P = 0.13; Table 7) or with distance from the Galápagos Islands (chi-square tests for spring and autumn: P = 0.8 and 0.13, respectively).

Band-rumped Storm-Petrel

Pelagic distribution.—We observed this species in two disjunct sections of the eastern Pacific Ocean: a relatively large concentration centered among the Galápagos Islands, and a few birds south of Hawaii. Crossin (1974) reported similar patterns. We observed the latter group only during autumn. We assumed that the two groups represent different breeding populations: the first from the Galápagos Islands (hereafter "Galápagos Band-rumped"), and the second probably from Hawaii or Japan (hereafter "Hawaiian–Japanese Band-rumped"). Owing to the strong relationship between density and distance to colony (see below), it is most likely that these latter birds were from Hawaii. We saw 990 Galápagos Band-rumped Storm-Petrels (adjusted = 902.6), and 10 of the probable Hawaiian–Japanese birds (adjusted = 9.7), not including 2 specimens collected and 9 other birds seen off transect in the western sector. Off the coast of the Americas, we observed this species from 24.80°N to 23.27°S; we did not observe them >1,807 km from the mainland. Therefore, our analyses for this species included surveys in waters from 25°N to 25°S, and offshore to 180°W (Table 1).

During both austral spring and autumn, Galápagos Band-rumped Storm-Petrels were congregated around the Galápagos Islands, though individuals were spread farther to the north and south during austral autumn (Fig. 15). This species was found primarily over the Cocos Ridge and Galápagos Fracture Zone, where depths are mostly consistent with those of the continental slope. Not surprisingly, Band-rumped Storm-Petrels were significantly more abundant over the continental slope than over shelf and pelagic waters (Sidak tests, both P < 0.001; Fig. 4H).

During both spring and autumn, densities of Band-rumped Storm-Petrels were highest in the Costa Rica Current and were significantly lower in the northern Humboldt Current, central Humboldt Current, and SEC (Sidak tests, all P < 0.001; Fig. 16A). Densities were also higher in both seasons in the northern Humboldt Current than in the central Humboldt Current (P < 0.001), though density differed little between the northern Humboldt Current and the SEC (both P > 0.3).

Habitat use.—The multiple-regression model for Band-rumped Storm-Petrels explained 27% of the variation in their at-sea distribution (Table 3), though the high variance explained was largely attributable to the very strong relationship between density and colony distance. Densities of Band-rumped Storm-Petrels increased with sea-surface temperature and decreased with increase in thermocline depth, wind speed, and distance from land and colony. The quadratic relationship between density and sea-surface temperature reflects a density increase to temperatures of 25°C, followed by a density decline with further temperature increase. The quadratic effect of thermocline

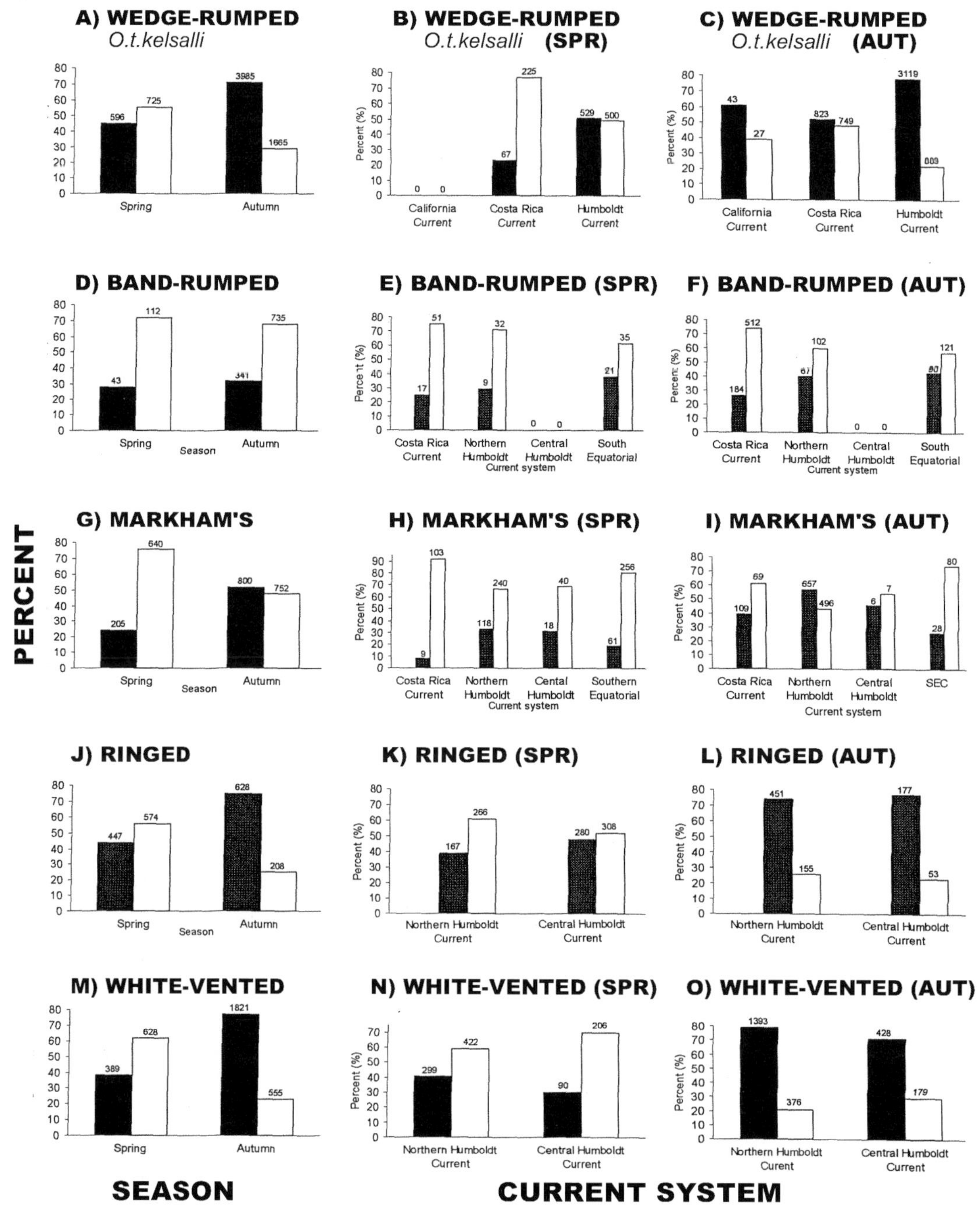

FIG. 14. Allocation of behavior among five storm-petrel taxa in different seasons and current systems, 1980–1995 (feeding–resting = light bar; in transit = dark bar; see Figure 5 for definitions of current-system acronyms). Values above bars are the adjusted numbers of birds observed (rounded off to nearest whole number).

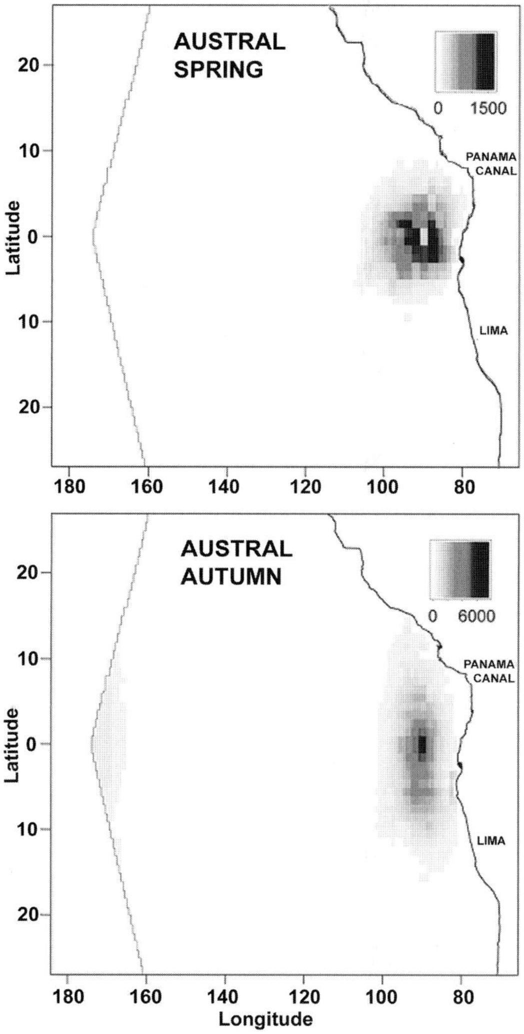

FIG. 15. Predicted distribution (birds per 1° latitude × 1° longitude cell) of Band-rumped Storm-Petrels during austral spring and autumn, 1980–1995. See Figure 3 for further details.

depth resulted from a decrease in density as thermocline depth increased to 75 m, followed by leveling in density where thermocline depth was greater. The quadratic effect of colony distance reflects a decline in density as distance increased to 2,500 km, followed by density leveling at greater distances.

The significant interactions between the effects of season and mainland distance as well as colony distance were both attributable to a greater decrease in density with distance in austral spring than in autumn (Table 4).

Pelagic behavior.—The proportion of Band-rumped Storm-Petrels that were feeding–resting (vs. flying in transit) did not vary significantly between seasons (χ^2 = 0.98, df = 1, P = 0.3; Fig. 14D). During austral spring, the proportion of Band-rumped Storm-Petrels observed foraging–resting differed little among the Costa Rica Current, northern Humboldt Current, and SEC (χ^2 = 2.30, df = 2, P < 0.001; Fig. 14E; central Humboldt Current excluded, because no Band-rumped Storm-Petrels were observed there in spring). During autumn, behavior differed significantly among the three areas (χ^2 = 25.53, df = 2, P < 0.001; Fig. 14F; central Humboldt Current excluded, because no Band-rumped Storm-Petrels were observed there in autumn), primarily because of a greater tendency for birds to feed when in the SEC and northern Humboldt Current than in the Costa Rica Current.

Population estimates.—Selected models for the distribution and abundance of Galápagos Band-rumped Storm-Petrels included all independent variables during spring, and all except latitude and distance to mainland in autumn (Table 5). Variables were modeled as smoothes, except for distance to colony in autumn.

Coefficients of variation for the spring and autumn abundance estimates of Band-rumped Storm-Petrels were about 11 and 15, respectively, which indicates that the GAMs were successful in fitting the survey data (Table 6). Our abundance estimates of Band-rumped Storm-Petrels during spring and autumn were 122,700 and 475,700, respectively (95% CI: 95,400–140,000 in spring and 340,500–600,300 in autumn). Thus, abundance was ~3.9× greater during autumn than in spring.

The number of Band-rumped Storm-Petrels seen in the western sector was too small to attempt a population estimate. Censuses to the north and west, especially in the vicinity of Hawaii, are needed to assess the size of this population. Crossin (1974), from a study during the 1960s, reported substantially more Band-rumped Storm-Petrels near the Hawaiian Islands than we found.

Markham's Storm-Petrel

Pelagic distribution.—We recorded 2,138 Markham's Storm-Petrels (adjusted = 1,923.7) between 16.55°N and 29.90°S and west along the equator to 118.02°W. This extends the

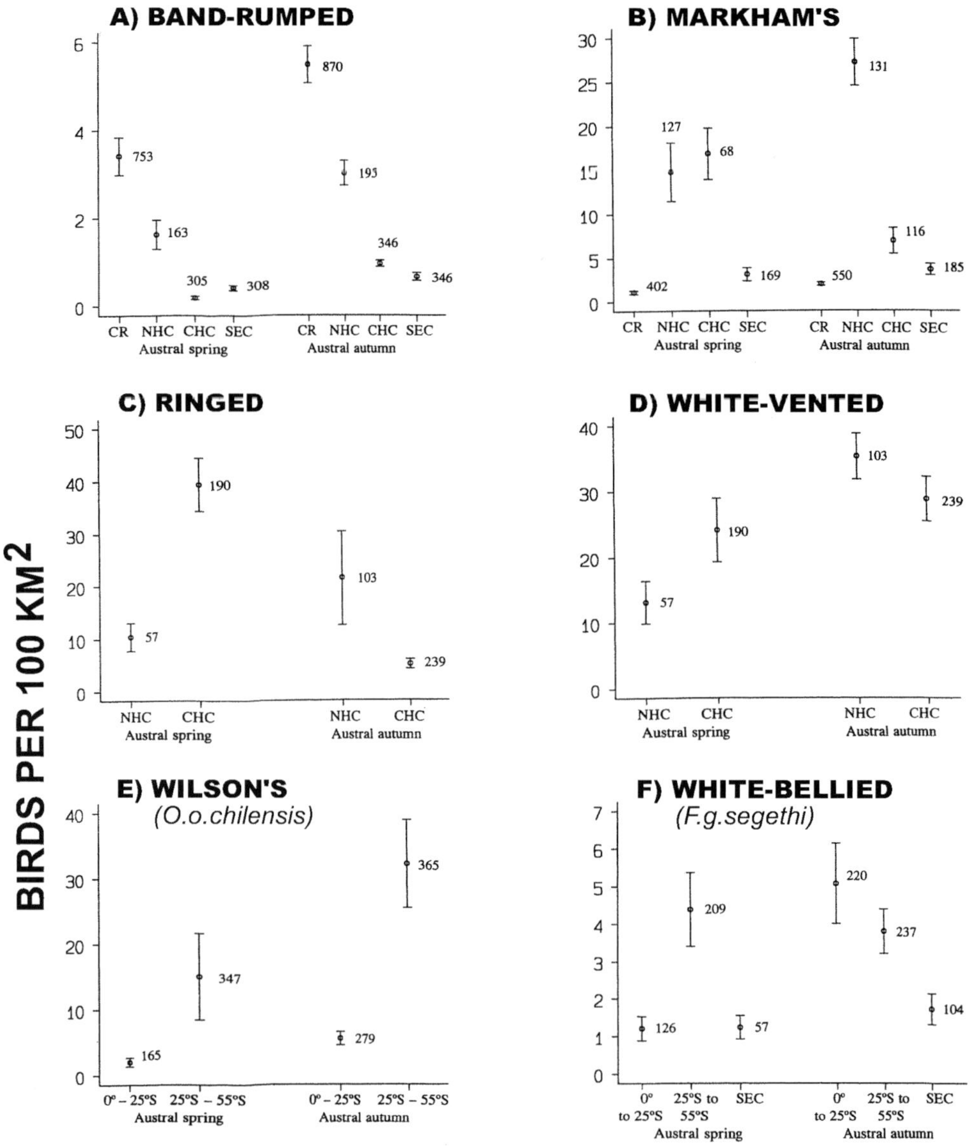

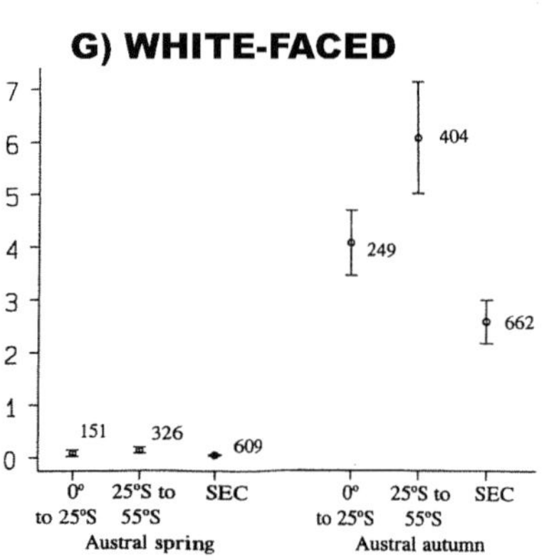

FIG. 16. Densities (birds per 100 km^2 of ocean surface; means ± SE) of seven storm-petrel taxa as related to current systems within the pelagic range of each taxon, 1980–1995 (see Table 2). See Figure 5 for definitions of current-system acronyms. Numbers adjacent to means denote number of survey segments.

contiguous range reported for the west (Crossin 1974) but is similar to that reported for the south (Murphy 1936). Therefore, analyses for this species pertain to waters from 18°N to 30°S and west to 115°W (Table 1). During austral autumn (the nonbreeding season), densities of Markham's Storm-Petrels were highest just offshore of central Peru (Fig. 17). Their distribution changed markedly in spring, at which time it became bifurcated, with large numbers concentrated just offshore of southern Peru and northern Chile and another high concentration centered ~1,700 km west of that area. Few were seen in these zones anywhere off the Americas in autumn. Densities of Markham's Storm-Petrels were significantly greater in continental shelf than in pelagic waters (Sidak test, $P < 0.001$; Fig. 4I); densities did not differ between shelf and slope waters ($P > 0.5$) or between slope and pelagic waters ($P > 0.1$). The proportion of adults versus subadults was significantly greater among Markham's Storm-Petrels collected within 200 km of shore than in those collected at distances >500 km, where subadults predominated ($\chi^2 = 8.04$, df = 1, $P < 0.01$; Table 7).

During both austral spring and autumn, density of Markham's Storm-Petrels was higher in the northern and central Humboldt Current than in the Costa Rica Current and SEC (Sidák tests, all $P < 0.001$; Fig. 16B); during both seasons, density was also higher in the SEC than in the Costa Rica Current (both $P < 0.025$).

Additional information on annual cycle.—We collected a female Markham's Storm-Petrel off Peru on 21 July 1987. It had a regressed follicle 7 mm in diameter, which indicated that she had recently laid an egg. Two other females collected in July had enlarged follicles 3.5 and 4.0 mm in diameter, respectively. Information on egg formation in White-faced Storm-Petrels indicates that a follicle 3.5 mm in width represents 2–3 days of rapid yolk-formation and that the eggs being formed would have been laid in ~14 days (Astheimer and Grau 1990, C. R. Grau pers. comm.). These data extend the known egg-laying dates for the species from mid-June (Murphy 1936) to mid-July through early August.

Habitat use.—The multiple-regression model for Markham's Storm-Petrels explained 21% of the variation in their at-sea distribution (Table 3). Densities of this taxon increased with sea-surface salinity and decreased with increase in sea-surface temperature and mainland distance. The quadratic relationship between density and sea-surface temperature reflects a decrease in density associated with temperatures increasing from 12°C to 23°C, followed by a decline associated with higher temperatures. The quadratic relationship between density and sea-surface salinity reflects stable density associated with salinities of 32 to 33.5 ppt, followed by a density increase with increased salinity. The quadratic effect of wind speed reflects stable

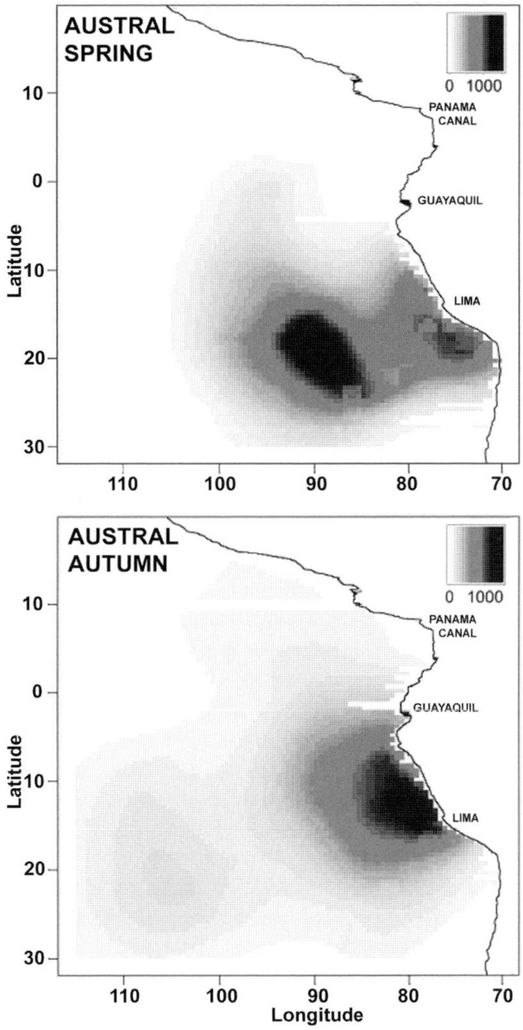

FIG. 17. Predicted distribution (birds per 0.5° latitude × 0.5° longitude cell) of Markham's Storm-Petrels during austral spring and autumn, 1980–1995. See Figure 3 for further details.

density with winds of 0–25 km h^{-1}, followed by density decrease with higher wind speeds.

The significant interaction between the effect of season and sea-surface salinity reflects a stronger increase in density with salinity during austral spring than in autumn (Table 4). The interaction of season with thermocline depth results from a strong decrease in density with increase in thermocline depth in spring, compared with no effect of thermocline depth on density during autumn.

Pelagic behavior.—The proportion of Markham's Storm-Petrels that was feeding–resting (vs. flying in transit) was significantly higher during austral autumn than in spring ($\chi^2 = 118.11$, df = 1, $P < 0.001$; Fig. 14G). During austral spring, behaviors differed significantly among the four ocean areas including the Costa Rica Current, the northern and central Humboldt Current, and the SEC ($\chi^2 = 36.46$, df = 3, $P < 0.001$; Fig. 14H), because of higher proportions of feeding–resting birds in the northern and central Humboldt Current than in the Costa Rica Current or SEC. During austral spring, behaviors also differed significantly among the four ocean areas ($\chi^2 = 59.12$, df = 3, $P < 0.001$; Fig. 14I), again primarily because of higher proportions of feeding–resting birds in the northern and central Humboldt Current than in the Costa Rica Current or SEC.

Population estimates.—Selected models for the distribution and abundance of Markham's Storm-Petrels included all independent variables in both seasons, except for longitude during spring (Table 5); all variables were modeled as smoothes except ocean depth, which was modeled as linear during austral spring.

Coefficients of variation for the austral spring and autumn abundance estimates of Markham's were 12, which indicates that the GAMs were very successful in fitting the survey data (Table 6). Our abundance estimates of Markham's during austral spring and autumn were 806,500 and 1,100,000, respectively (95% CI: 700,000–1,000,000 in spring and 1,000,000–1,600,000 in autumn). Thus, abundance was ~1.4× greater in autumn than in spring.

Although sample sizes were problematic, ratios of adults to subadults did not vary significantly between seasons ($\chi^2 = 2.32$, df = 1, $P = 0.13$; Table 7) or with distance from the Galápagos Islands (chi-square tests for spring and autumn: $P = 0.8$ and 0.13, respectively).

Ringed Storm-Petrel

Pelagic distribution.—We observed this species only in the Humboldt Current, recording 1,866 individuals (adjusted = 1,756.8) between 3.73°S and 31.48°S and to 481 km offshore. Therefore, analyses for this species pertain to waters from 3°S to 32°S and to 500 km offshore (Table 1). During austral spring, the species was concentrated in waters off southern Ecuador and northern Peru; during autumn, the largest concentrations were observed offshore of Paracas, Peru, to waters off southern Peru and northern Chile (Fig. 18). Density of Ringed Storm-Petrels was significantly higher over continental shelf waters than over slope or pelagic waters (Sidak tests, both $P < 0.025$; Fig. 19A) and significantly higher over pelagic than over slope waters ($P < 0.001$). During austral autumn, density was significantly higher in the central than in the northern Humboldt Current (*t*-test, $P < 0.001$; Fig. 16C); but in spring, density was significantly higher in the northern Humboldt Current (*t*-test, $P < 0.01$).

Additional information on annual cycle.—A female we collected on 24 November 1986 had an enlarged follicle (4.5 mm), a finding that provided the first evidence for an austral spring–summer breeding season in this species (see above; Murphy 1936). Information on egg-formation in White-faced Storm-Petrels indicates that the egg being formed would have been laid in ~14 days (i.e., early December) (Astheimer and Grau 1990, C.R. Grau pers. comm.).

Habitat use.—The multiple-regression model for Ringed Storm-Petrels explained 20% of the variation in their at-sea distribution (Table 3). Densities of this taxon increased with sea-surface salinity and decreased with increase in sea-surface temperature and mainland distance. The quadratic relationship between density and sea-surface temperature reflects a decrease in density associated with temperatures increasing from 12°C to 23°C, followed by a decline associated with higher temperatures. The quadratic relationship between density and sea-surface salinity reflects stable density associated with salinities of 32–33.5 ppt, followed by a density increase in saltier waters. The quadratic effect of wind speed reflects stable density with winds of 0–25 km h^{-1}, followed by density decrease with higher wind speeds.

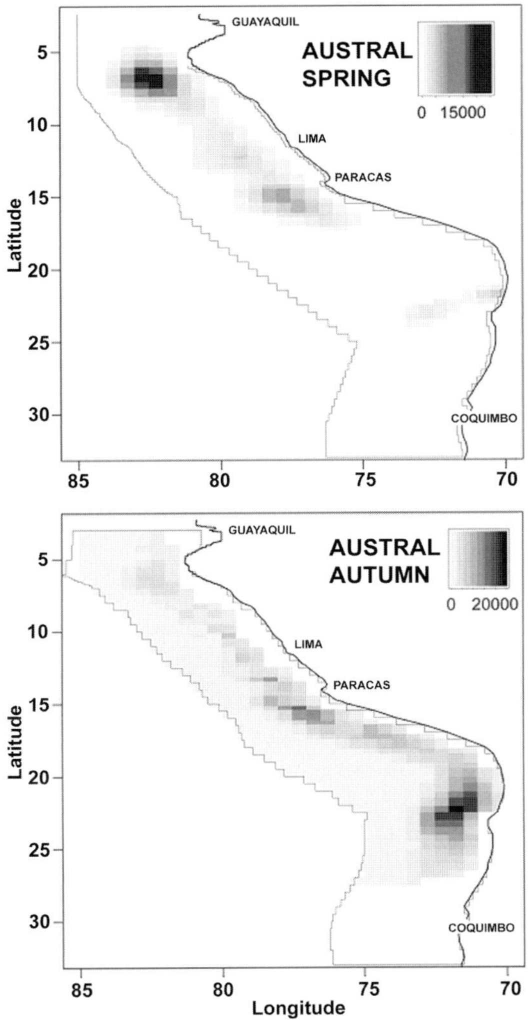

FIG. 18. Predicted distribution (birds per 0.5° latitude × 0.5° longitude cell) of Ringed Storm-Petrels during austral spring and autumn, 1980–1995. See Figure 3 for further details.

The significant interaction between the effect of season and sea-surface salinity reflects a stronger increase in density with salinity during austral spring than in autumn (Table 4). The interaction of season with thermocline depth reflects a strong decrease in density with increase in thermocline depth in spring, compared with no effect of thermocline depth on density during autumn.

Pelagic behavior.—The proportion of Ringed Storm-Petrels that was feeding–resting (vs. flying in transit) was significantly higher during austral autumn than in spring ($\chi^2 = 185.18$, df = 1, $P < 0.001$; Fig. 14J). During austral spring, a significantly higher proportion of Ringed Storm-Petrels recorded in the central Humboldt Current was feeding–resting, compared with those observed in the northern Humboldt Current ($\chi^2 = 8.30$, df = 1, $P < 0.01$; Fig. 14K). During austral spring, Ringed Storm-Petrel behavior in the two sections of the Humboldt Current was similar ($\chi^2 = 0.57$, df = 1, $P = 0.4$; Fig. 14L).

Population estimates.—Selected models for the distribution and abundance of Ringed Storm-Petrels included all independent variables in both seasons, except for ocean depth during spring (Table 5).

Coefficients of variation for the austral spring and autumn abundance estimates of Ringed Storm-Petrel during austral spring and autumn were about 18 and 15, which indicates that the GAMs were moderately successful in fitting the survey data (Table 6). Our abundance estimates during spring and autumn were 637,200 and 1,011,900, respectively (95% CI: 0.5–0.8 million in spring and 900,000–1,500,000 in autumn). Thus, abundance was ~1.6× greater in autumn than in spring.

White-vented Storm-Petrel

Pelagic distribution.—We observed 3,054 White-vented Storm-Petrels, only in the Humboldt Current (adjusted = 2,935.9) and at distances >100 km from the Galápagos Islands; we recorded four within 100 km of the Galápagos Islands but excluded those from the following because they were likely of the race *O. gracilis galapagoensis*). We saw no White-vented Storm-Petrels in a 550-km gap between the Galápagos and offshore Ecuador; all but the four recorded adjacent to the Galápagos were between 3.25° and 31.48°S, within 475 km of the coast of South America. Similar patterns of occurrence were observed by Loomis (1918), Murphy (1936), and Crossin (1974). Therefore, our analyses for this species pertain to waters from 3°N to 32°S and to 500 km offshore (Table 1).

During the austral spring breeding season, White-vented Storm-Petrels were concentrated just offshore of northern to central Peru (Fig. 20). During autumn, the highest concentration extended farther offshore between southern Ecuador and northern Chile. Densities of White-vented Storm-Petrels were significantly higher

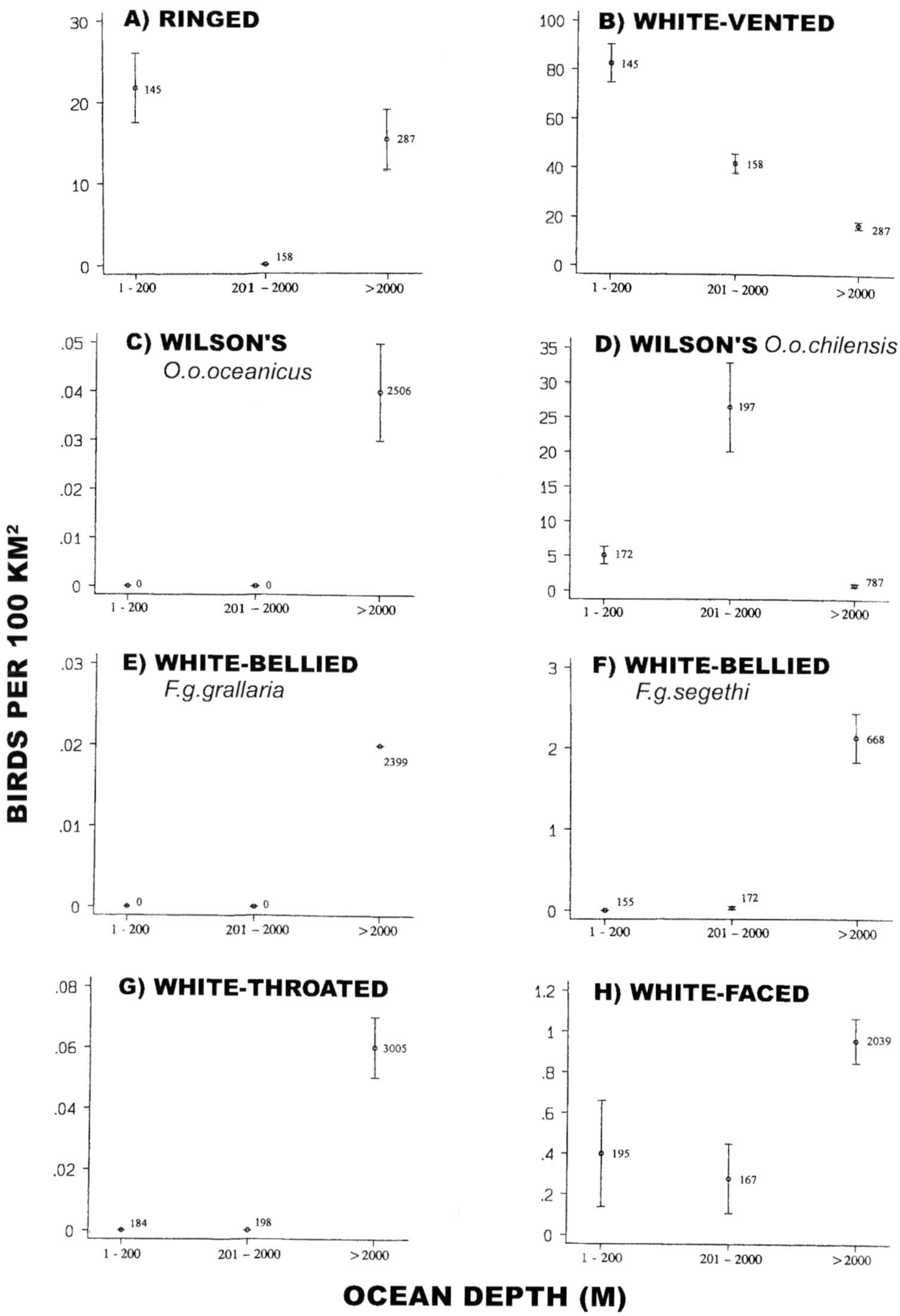

FIG. 19. Densities (birds per 100 km^2 of ocean surface; means ± SE) of eight storm-petrel taxa as related to ocean depth within the pelagic range of each, 1980–1995 (see Table 2). Depth 1–200 m = continental shelf; 201–200 m = continental slope; >2,000 m = pelagic waters. Numbers adjacent to means denote number of survey segments.

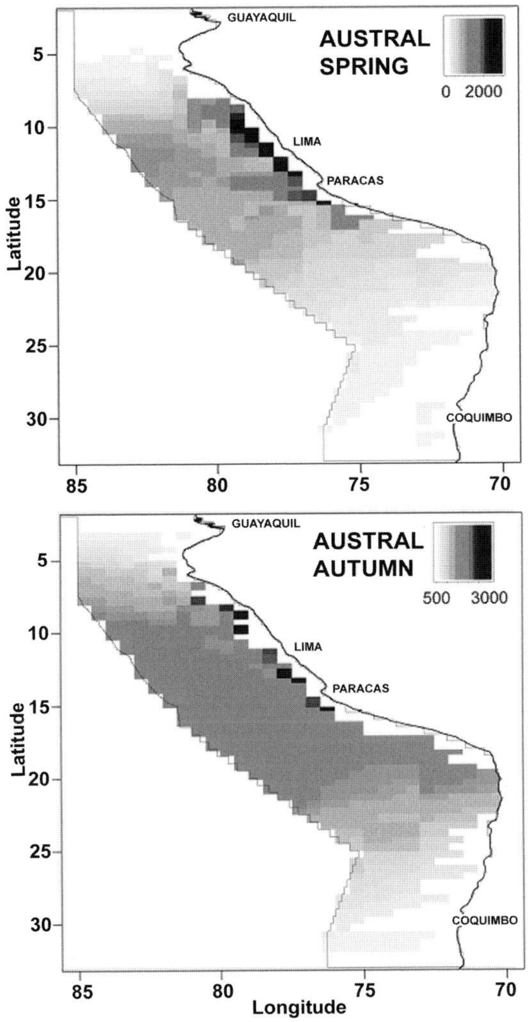

FIG. 20. Predicted distribution (birds per 0.5° latitude × 0.5° longitude cell) of White-vented Storm-Petrels during austral spring and autumn, 1980–1995. See Figure 3 for further details.

over the continental shelf than over slope or pelagic waters (Sidak tests, both $P < 0.001$; Fig. 19B) and were significantly higher over slope than over pelagic waters ($P < 0.001$). During austral spring, density of White-vented Storm-Petrels was higher in the central than in the northern Humboldt Current (t-test, $P < 0.001$; Fig. 16D); during autumn, density was similar in the two areas ($P > 0.5$).

Habitat use.—The multiple-regression model for White-vented Storm-Petrels explained 57% of the variation in their at-sea distribution (Table 3). Densities of this taxon decreased with increase in sea-surface temperature, thermocline strength, and mainland distance. The quadratic relationship between density and sea-surface temperature reflects stable densities associated with temperatures of 11–15°C, followed by a decline associated with higher temperatures. The quadratic relationship between density and thermocline strength reflects stable density associated with slope strength of 1–5°C change, followed by an increase in density with more stratified thermoclines. The quadratic effect of mainland distance reflects a decline in density as distance increased to 300 km, followed by a leveling of density at distances to 500 km.

The significant interaction between the effect of season and sea-surface temperature reflects a stronger increase in density with temperature during austral spring than in autumn (Table 4). The interaction of season with sea-surface salinity reflects a strong decrease in density with increase in salinity in autumn, compared with no effect of thermocline depth on density during spring.

Pelagic behavior.—The proportion of White-vented Storm-Petrels that was feeding–resting (vs. flying in transit) was significantly higher during austral autumn than in spring (χ^2 = 462.22, df = 1, $P < 0.001$; Fig. 14M). During both austral spring and autumn, a significantly higher proportion of these petrels were feeding in the northern than in the central Humboldt Current (spring: χ^2 = 10.88, df = 1, $P < 0.001$; Fig. 14N; autumn: χ^2 = 17.12, df = 1, $P < 0.001$; Fig. 14O).

Population estimates.—Selected models for the distribution and abundance of White-vented Storm-Petrels included all independent variables in both seasons, with the exceptions of longitude and ocean depth during spring and longitude during autumn (Table 5).

Coefficients of variation for the austral spring and autumn abundance estimates of White-vented Storm-Petrels during austral spring and autumn were about 13 and 7, which indicates that the GAMs were very successful in fitting the survey data (Table 6). Our abundance estimates of White-vented Storm-Petrels during spring and autumn were 396,400 and 858,700, respectively (95% CI: 343–473,000 in spring and 811,000–1,026,000 in autumn). Thus, abundance was ~2.2× greater in austral autumn than in spring.

Wilson's Storm-Petrel (*Oceanites oceanicus oceanicus*)

Pelagic distribution.—Like Crossin (1974), we recorded two disjunct groups of this species, one in the central Pacific and the other confined to the Humboldt Current. We observed no Wilson's Storm-Petrels between 509 km offshore of South America and 119°W, a distance of ~3,000 km.

Individuals that we collected off Ecuador and Peru were markedly and consistently smaller than those we collected in the central Pacific (Appendix 1). Furthermore, measurements of Wilson's Storm-Petrels we collected off South America were consistent with those of Wilson's Storm-Petrels collected by others on Chilean breeding grounds, whereas those of birds we collected in the central Pacific were consistent with Wilson's Storm-Petrels collected by others on Antarctic breeding grounds. Therefore, we assume that all the Wilson's Storm-Petrels we saw in the central Pacific were of Antarctic origin (*O. o. oceanicus*) and that those encountered off South America were of Chilean origin (*O. o. chilensis*; for similar conclusions, see Murphy 1936, Crossin 1974). Thus, the distribution was largely the inverse of, but analogous to, that of Leach's Storm-Petrels from the north, with coastal forms (indicated by dark-rumped "markers") occurring in the easternmost portion of the study area.

We recorded only 22 individuals (adjusted = 20.2) representing *O. o. oceanicus*. These birds were confined to the SEC and scattered widely between 12.87°N and 2.65°S and between 119°W and 166°W (Fig. 21). All were in pelagic waters (Fig. 19C). Hence, analyses for these birds were confined to the region from 15°N to 5°S and from 300 km offshore to 170°W (Table 1). This taxon was too rare in the eastern Pacific Ocean to allow analyses of seasonal patterns and, therefore, we grouped the data for austral spring and autumn. The pelagic range of this form in the eastern Pacific Ocean is likely more extensive than is indicated by the range we observed (see Crossin 1974). Adults contributed the majority of *O. o. oceanicus* specimens collected in the central Pacific (Table 7).

Habitat use.—The multiple-regression model for *O. o. oceanicus* (seasons grouped) was not significant (Table 3).

Pelagic behavior.—A total of 73% (n = 19) of the 26 *O. o. oceanicus* recorded were feeding–resting; the remainder were flying in transit.

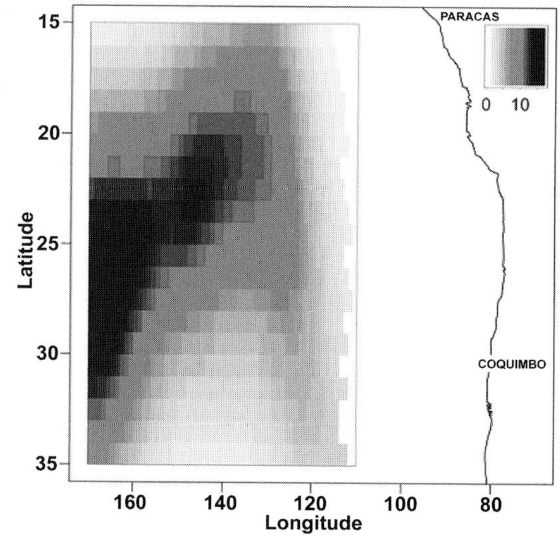

FIG. 21. Predicted distribution (birds per 1° latitude × 1° longitude cell) of the *O. o. oceanicus* form of Wilson's Storm-Petrel, seasons combined, 1980–1995. See Figure 3 for further details.

Population estimates.—The model selected for the distribution and abundance of *O. o. oceanicus* included all independent variables except depth (Table 5). The coefficient of variation for the abundance estimate of *O. o. oceanicus* was 22.4, which indicates that the GAMs were only moderately successful in fitting the survey data (Table 6). The abundance estimate was 8,054 (95% CI: 6,380–12,590).

Wilson's Storm-Petrel (*Oceanites oceanicus chilensis*)

Pelagic distribution.—We recorded 1,301 *O. o. chilensis* (adjusted = 1,263.5) within 509 km of the South American coast and between 3.10°S and 53.63°S. Similarly, Beck (in Murphy 1936) found them only to 250 km off Chile. Our analyses for this taxon pertain to waters from 3°S to 54°S and to 550 km offshore (Table 1). During the austral spring nonbreeding season, this taxon was most concentrated off southern Chile, though high concentrations extended along the coast to northern Chile (Fig. 22). During the austral autumn breeding season, their concentrations were bifurcated, with areas of high abundance off central and southern Chile. Density of *O. o. chilensis* was significantly higher over the continental slope than over shelf or pelagic waters (Sidak tests, both $P < 0.001$; Fig. 19D), though

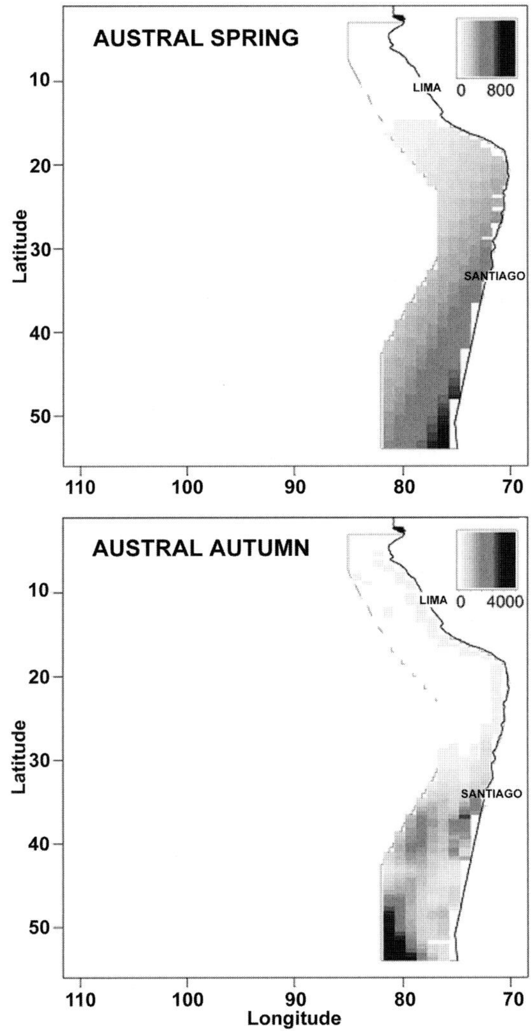

FIG. 22. Predicted distribution (birds per 0.5° latitude × 0.5° longitude cell) of the *O. o. chilensis* form of Wilson's Storm-Petrel during austral spring and autumn, 1980–1995. See Figure 3 for further details.

density over the shelf was higher than over pelagic waters ($P < 0.001$). Subadults predominated in a sample of *O. o. chilensis* collected off Ecuador (Table 7), apparently the northern extreme of their regular range.

Habitat use.—The multiple-regression model for *O. o. chilensis* explained 26% of the variation in this form's at-sea distribution (Table 3). Densities of this taxon decreased with increase in sea-surface temperature and wind speed, but increased with sea-surface salinity. The quadratic relationship between density and sea-surface temperature results from a decrease in density as temperature increases from 8°C to 15°C, followed by an accelerating decline in density as temperature increases further. The quadratic relationship between density and wind speed reflects density decrease as wind speed increased from 0 to 20 km h^{-1}, followed by an even greater decrease as wind speed increased further.

The significant interaction between the effect of season and sea-surface temperature reflects a stronger increase in density with temperature during austral autumn than in spring (Table 4). The interaction of season with wind speed reflects a strong decrease in density with increase in wind speed in autumn, compared with no effect of wind speed on density during spring.

Pelagic behavior.—A significantly higher proportion of *O. o. chilensis* was feeding–resting (vs. flying in transit) during austral autumn than in spring ($\chi^2 = 50.80$, df = 1, $P < 0.001$; Fig. 23A). During both austral spring and autumn, a significantly higher proportion of this race was feeding in the northern than in the central Humboldt Current (spring: $\chi^2 = 13.23$, df = 1, $P < 0.001$; Fig. 23B; autumn: $\chi^2 = 23.65$, df = 1, $P < 0.001$; Fig. 23C).

Population estimates.—Selected models for the distribution and abundance of *O. o. chilensis* included only the linear effects of latitude and longitude in austral spring, and all independent variables except for longitude during autumn (Table 5).

Coefficients of variation for estimates of *O. o. chilensis* abundance during austral spring and autumn were about 14 and 18, which indicates that the GAMs were moderately successful in fitting the survey data (Table 6). Our abundance estimates of *O. o. chilensis* during spring and autumn were 866,600 and 406,000, respectively (95% CI: 782,000–1,154,000 in spring and 295,000–563,000 in autumn). Thus, abundance was ~2.1× greater in austral spring than in autumn.

This estimate is undoubtedly low, because we did not census waters south of 50°S, where these birds may be even more abundant (Murphy 1936).

White-bellied Storm-Petrel

Mensural differences among White-bellied Storm-Petrels we collected in the eastern Pacific Ocean indicate that three races were present

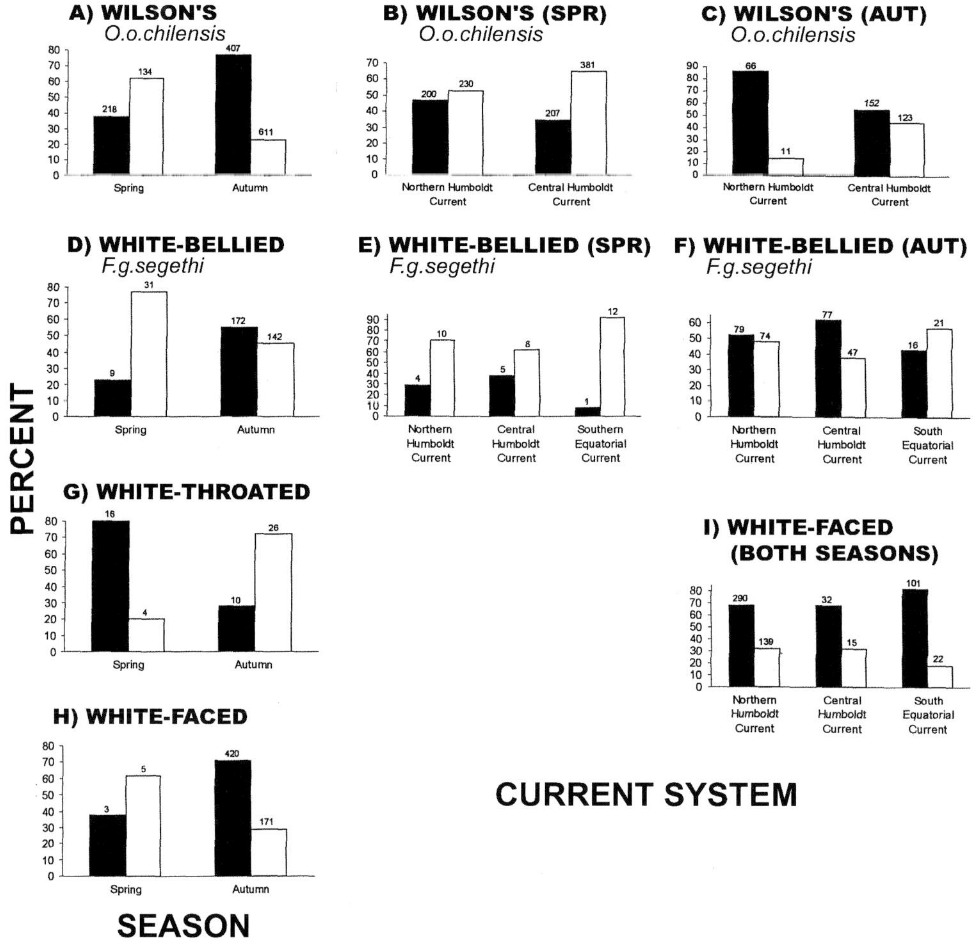

FIG. 23. Allocation of behavior of four storm-petrel taxa in different seasons and current systems, 1980–1995 (feeding–resting = light bar; in transit = dark bar). See Figure 5 for definitions of current acronyms. Values above bars are the adjusted numbers of birds observed (rounded to nearest whole number).

(Appendix 2). We collected one subadult of the largest race (*G. g. titan*) at 5°N, 140°W on 23 November 1988 (USNM 597204; Appendix 2), ~3,000 km north of the breeding location of this race on the Austral Islands. Before this record, their at-sea occurrence was (possibly) represented by a single bird collected 450 km south of the Galápagos Islands (Murphy 1936).

The morphology of White-bellied Storm-Petrels collected within or immediately adjacent to the Humboldt Current matched well with that of the smallest form, *F. g. segethi*, which breeds on the Juan Fernandez Islands off central Chile (Appendix 2). On the other hand, measurements of the 15 birds we collected between 120°W and 150°W (Appendix 2) are larger than *F. g. segethi* and smaller than *F. g. titan*, but cannot be separated mensurally from other intermediate-sized races, such as the "*F. g. lineata*" bird from near the Marquesas, *F. g. grallaria* from Lord Howe and Kermadec islands, or *F. g. tristanensis* from the southern Atlantic Ocean (Appendix 2). We did not collect or see any melanistic White-bellied Storm-Petrels. Another clue to identity appears to be molt. All but one were collected during austral spring (late October–late November). Each of 17 we collected was undergoing body molt and was just finishing or had just completed its primary molt. This molt chronology indicates an austral summer–autumn breeding season consistent with the breeding period known for both of the medium-sized races: *F. g. grallaria* and *F. g. leucogaster*. Given the breeding location of the latter in the southern Atlantic Ocean, the equatorial White-bellied Storm-Petrels are likely not of that race. Therefore, we assume here that they represent *F. g. grallaria*. Because of the 3,500-km hiatus between forms recorded in the west and east, we have assumed that the latter, much smaller, form is *F. g. segethi*.

Fregetta grallaria grallaria

Pelagic distribution. We recorded 16 White-bellied Storm-Petrels (adjusted = 12.5) between 3.82°N and 13.27°S in the equatorial region that, visually, were smaller than *F. g. titan* (adjusted = 15.1). During surveys, we observed no birds that we suspected to be *F. g. titan*. During austral autumn, *F. g. grallaria* was sparsely distributed in pelagic waters (Fig. 19E) of the SEC from 119°W to 175°W. We saw only one during austral spring and have grouped seasons for analyses of abundance, distribution, and habitat use; those analyses pertain to waters from 4°N to 20°S and from 110°W to 175°W. These birds were most abundant west of 140°W (Fig. 24). Seventy-six percent (n = 17) of the *F. g. grallaria*-like birds we collected in the equatorial Pacific were adults; the single bird collected in austral spring was a subadult (Table 7).

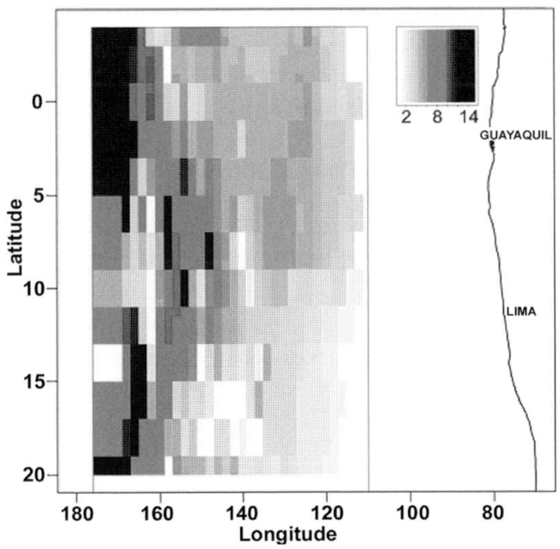

FIG. 24. Predicted distribution (birds per 0.5° latitude × 0.5° longitude cell) of the *F. g. grallaria* form of White-bellied Storm-Petrel, seasons combined, 1980–1995. See Figure 3 for further details.

Habitat use.—The multiple-regression model for *F. g. grallaria*-like birds explained only 2% of the variation in their at-sea distribution (Table 3). Densities of this taxon decreased with increase in sea-surface temperature but increased with sea-surface salinity and thermocline strength. The significant interaction between the effect of season and sea-surface salinity reflects a significant increase in density with salinity in austral spring, compared with no effect of salinity in autumn (Table 4). The interaction of season with thermocline strength reflects a significant increase in density with thermocline strength in austral spring, compared with no effect of thermocline strength in autumn.

Pelagic behavior.—A total of 56% (n = 9) of the 16 *F. g. grallaria* recorded was feeding–resting; the others were flying in transit.

Population estimates.—Selected models for the distribution and abundance of *F. g. grallaria*-like birds included only the linear effects of latitude and longitude in austral spring, and all independent variables except for longitude during autumn (Table 5).

The coefficient of variation for the abundance estimate (seasons grouped) of *F. g. grallaria*-like birds was ~26, which indicates that the GAMs did not fit the survey data very well (Table 6). Our abundance estimate for *F. g. grallaria* is 2,866 (95% CI: 2,198–5,713).

Fregetta grallaria segethi

Pelagic distribution.—We saw 406 (adjusted = 353.5) birds of this form between 0.65°N and 35.87°S and within 1,700 km of the South American coast; analyses for this taxon pertain to waters from 4°N to 36°S and from the mainland to 100°W (Table 1). Like *F. g. grallaria* (see above), these birds were almost exclusively pelagic (Fig. 19F), given that we observed only one (<0.5%) inside the 2,000-m isobath. During both spring and fall, highest densities were observed off northern Chile (Fig. 25); during austral autumn, though, their distribution bifurcated, with a second concentration appearing well offshore of Peru.

Habitat use.—The multiple-regression model for *F. g. segethi* explained 40% of the variation in their at-sea distribution (Table 3). Densities of this taxon decreased with increase in sea-surface temperature and distance from the colony at the Juan Fernandez Islands but increased with

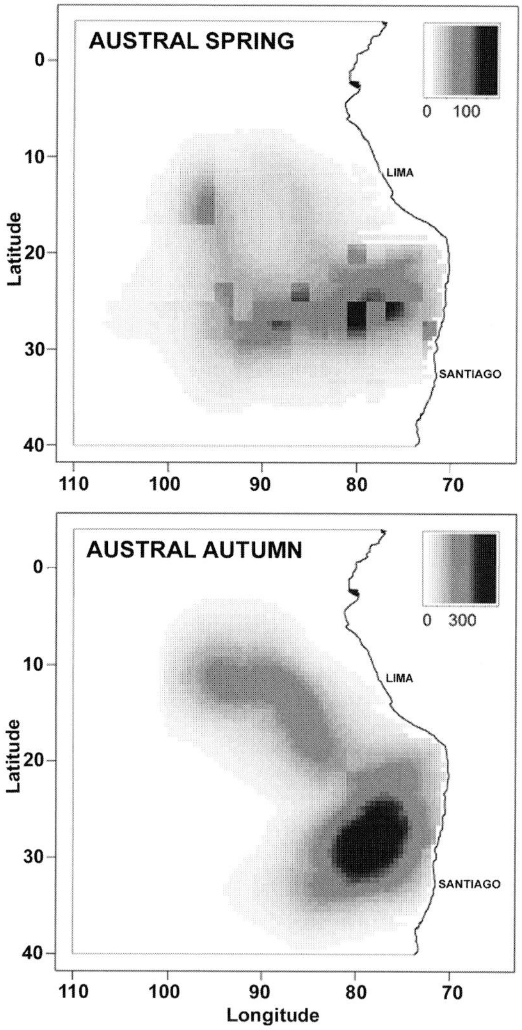

FIG. 25. Predicted distribution (birds per 0.5° latitude × 0.5° longitude cell) of the *F. g. segethi* form of White-bellied Storm-Petrel during austral spring and autumn, 1980–1995. See Figure 3 for further details.

thermocline depth and distance from the mainland. The quadratic relationship between density and sea-surface temperature reflects a stable density associated with temperatures of 8–15°C, followed by a decrease at higher temperatures. The quadratic relationship between density and mainland distance reflects a density increase from 0 to 1,000 km, followed by leveling at greater distances. The quadratic relationship between density and colony distance reflects a density decrease as distance increased from 0 to 1,200 km, followed by leveling at greater distances.

The significant interactions between the effects of season, sea-surface temperature, and colony distance reflect stronger increases in density with decrease in temperature and distance during austral autumn than in spring (Table 4). The interaction of season with wind speed reflects a strong decrease in density with increase in wind speed in spring, compared with no effect of wind speed on density during autumn.

Pelagic behavior.—A significantly higher proportion of *F. g. segethi* was feeding–resting (vs. flying in transit) during austral autumn than in spring ($\chi^2 = 14.79$, df = 1, $P < 0.001$; Fig. 23D). During both austral spring and autumn, the proportion of birds feeding–resting (vs. flying in transit) did not differ among the northern and southern Humboldt Current and the SEC (spring: $\chi^2 = 3.43$, df = 2, $P > 0.2$; Fig. 23E; autumn: $\chi^2 = 5.28$, df = 2, $P = 0.07$; Fig. 23F), because of higher proportions feeding–resting in the central and northern Humboldt Current than in the SEC.

Population estimates.—Selected models for the distribution and abundance of *F. g. segethi* included all five covariates in both seasons; all were modeled as smoothes (Table 5).

Coefficients of variation for abundance estimates of *F. g. segethi* during austral spring and autumn were about 11 and 12, which indicates that the GAMs were successful in fitting the survey data (Table 6). Our abundance estimates of *F. g. segethi* during spring and autumn were 114,600 and 442,500, respectively (95% CI: 84,000–129,000 in spring and 309,000–545,000 in autumn). Thus, abundance was ~3.9× greater in austral autumn than in spring.

White-throated Storm-Petrel

Pelagic distribution.—We recorded 61 White-throated Storm-Petrels (adjusted = 56.1) between latitudes 11.50°N and 12.50°S and between 96.75°W and 172.10°W. The distribution was concentrated to the north of Easter Island and east of the Line Islands in austral spring, but was concentrated in the vicinity of the Marquesas during austral autumn (Fig. 26). These birds were exclusively pelagic (Fig. 19G). Adults predominated in specimens collected (Table 7).

Additional information on annual cycle.—We collected five specimens (three males and two females) with brood patches, enlarged gonads or distended oviducts, and fresh plumage. Interestingly, two of these birds were collected

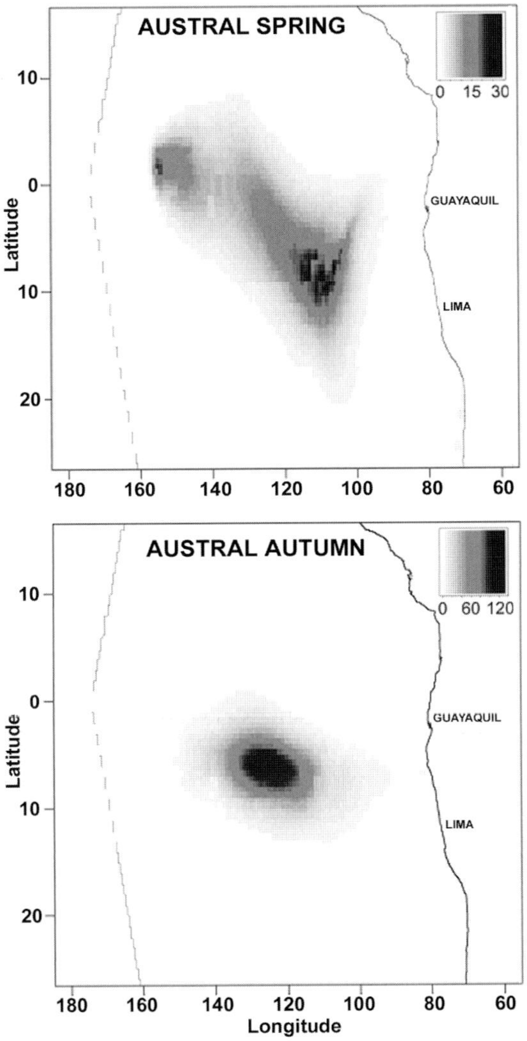

FIG. 26. Predicted distribution (birds per 1° latitude × 1° longitude cell) of White-throated Storm-Petrels during austral spring and autumn, 1980–1995. See Figure 3 for further details.

during late October and three were collected from late April to early May, all at 140°W: three on the equator, and two at 5°S. These locations are 875 and 325 km, respectively, from the nearest of the Marquesas Islands, and >2,000 km from Christmas Island, the nearest known breeding location. These observations are consistent with an extended or biannual breeding season and a suspected breeding population on the Marquesas (Thibault 1973).

Habitat use.—The multiple-regression model for White-throated Storm-Petrel explained only 5% of the variation in this form's at-sea distribution (Table 3). Densities of these birds increased with sea-surface salinity and thermocline depth. There were no interactions between season and oceanographic variables.

Pelagic behavior.—A significantly larger proportion of White-throated Storm-Petrels fed–rested during austral spring than in autumn ($\chi^2 = 14.10$, df = 1, $P < 0.001$; Fig. 23G). Numbers observed in the two seasons were too small to subdivide the data for analyses separated between the SEC and ECC (the two current systems where this taxon was observed).

Population estimates.—Selected models for the distribution and abundance of White-throated Storm-Petrels included all five covariates in spring, and all but mainland distance for autumn (Table 5).

Coefficients of variation for abundance estimates of White-throated Storm-Petrels during austral spring and autumn were about 18 and 16, which indicates that the GAMs were moderately successful in fitting the survey data (Table 6). Our abundance estimates for this species during spring and autumn were 52,300 and 20,800, respectively (95% CI: 36,000–88,000 in spring and 16,000–26,000 in autumn). Thus, estimated abundance was ~2.5× greater in austral spring than in autumn.

White-faced Storm-Petrel

Pelagic distribution.—We observed 618 individuals (adjusted = 591.4) during austral autumn, but only 4 in spring (adjusted = 7.5). We recorded this species from 8.15°N to 44.65°S and between the South American coast and 154.20°W, or in waters from the southern end of the Humboldt Current north and west along the northern boundary of the SEC. Our analyses for this species pertain to waters within that area (Table 1). This range is thus farther to the west than that observed by Crossin (1974). During austral autumn, White-faced Storm-Petrels were concentrated off Peru from 8°S to 22°S, with high densities extending to 3,000 km offshore in those latitudes (Fig. 27). In spring, the largest concentration occurred off southern Chile, though moderately high densities were also found off Ecuador and well offshore near the equator at about 100–120°W. Densities of White-faced Storm-Petrels were significantly higher over pelagic waters than over continental

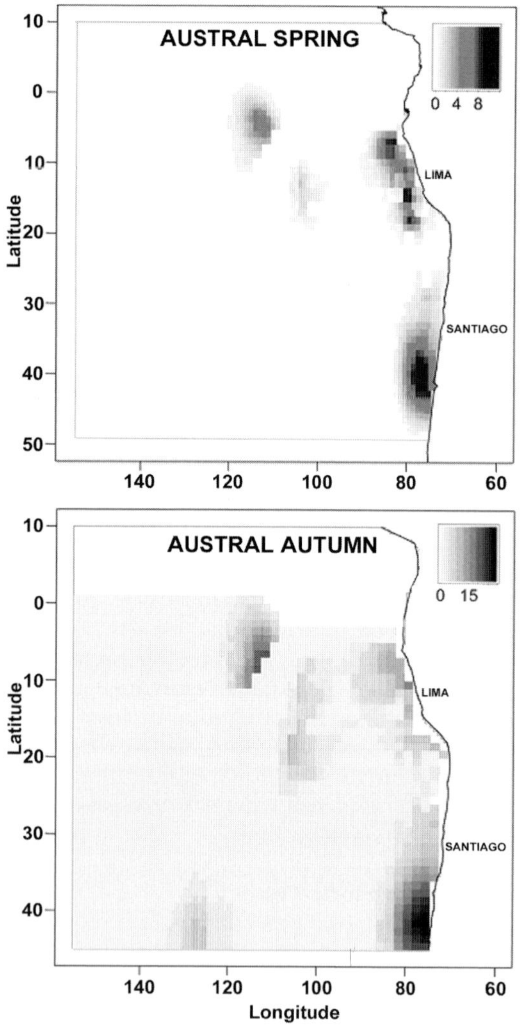

FIG. 27. Predicted distribution (birds per 1° latitude × 1° longitude cell) of White-faced Storm-Petrels during austral spring and autumn, 1980–1995. See Figure 3 for further details.

shelf or slope waters (Sidak tests, both $P < 0.001$; Fig. 19H); density differed little between the shelf and slope habitats ($P > 0.5$).

Interestingly, we recorded this species along the equator only in periods just preceding or during El Niño. In other years, we saw them only off South America. Adults predominated among individuals collected in the eastern tropical Pacific; proportions of adults to subadults did not vary between eastern and western sectors of the eastern Pacific Ocean ($\chi^2 = 3.28$, df = 1, $P = 0.07$; Table 7).

Origin of birds observed in the eastern Pacific Ocean.—As noted above, the origin of White-faced Storm-Petrels in our study area is likely the New Zealand (*maoriana*) population or the Australian *dulciae* population (Crossin 1974), or both. Although Murphy and Irving (1951) excluded *dulciae* as a possibility for one specimen collected off Ecuador (on the basis of color pattern and tail structure), they could not exclude the possibility that this bird, which appeared to be *maoriana*, represented the Tristan da Cunha *marina* population, because of the similar appearance of the two races. Murphy and Irving (1951) noted, however, that many species of seabirds originating from breeding grounds in New Zealand migrate to the Peru Current (see also Spear et al. 2003, 2005) and postulated a similar origin for *Pelagodroma* occurring in the eastern Pacific Ocean. Each of the 15 White-faced Storm-Petrels we collected also had characteristics of *maoriana–marina* (i.e., heavily marked neck pattern and markedly forked tails). It is unlikely that many of the birds we saw could have been *marina*, because of the low number or possible extermination of this race on Tristan da Cunha (see above).

Habitat use.—The multiple regression model for White-faced Storm-Petrels explained 14% of the variation in their at-sea distribution (Table 3). Densities of this taxon increased with sea-surface temperature, thermocline strength, mainland distance, and colony distance. The quadratic relationship between density and sea-surface salinity reflects stable density associated with salinities of 33–34.5 ppt and a decrease with higher salinities. The quadratic relationship between density and thermocline depth reflects density increase with thermocline depth from 0 to 75 m, followed by a decrease associated with a deeper thermocline. Similarly, the quadratic relationship between density and thermocline strength reflects increasing density as strength increased from an 8°C gradient, followed by decreasing density as thermocline strength intensified further. The quadratic relationship between density and mainland distance reflects an increase as distance increased from 0 to 1,200 km, followed by leveling at greater distances. The quadratic relationship between density and colony distance reflects density increase as distance increased from 0 to 5,000 km, followed by leveling at greater distances.

The significant interactions between the effects of season and sea-surface temperature, thermocline slope, and colony distance reflect significant increases in density with salinity, temperature, and distance during austral autumn, compared with no effects of these variables on density during spring (Table 4).

Pelagic behavior.—A significantly larger proportion of White-faced Storm-Petrels fed–rested during austral autumn than in spring ($\chi^2 = 4.29$, df = 1, $P < 0.04$; Fig. 23H). This comparison was problematic, however, because of low numbers recorded in spring. Numbers observed in spring were too small to subdivide by current system. When we grouped the two seasons, a significantly higher proportion of birds were feeding–resting (vs. flying in transit) when over the SEC than when over the northern or central Humboldt Current ($\chi^2 = 9.86$, df = 2, $P < 0.01$; Fig. 23I).

Population estimates.—Selected models for the distribution and abundance of White-faced Storm-Petrels included all covariates in both seasons, except longitude and colony distance (Table 5).

Coefficients of variation for abundance estimates of this species during austral spring and autumn were about 25 and 14, which indicates that the GAMs were successful in fitting the survey data in autumn but were only marginally so during spring (Table 6). Small numbers of birds observed were responsible for the lack of precision in spring. Our abundance estimates for White-faced Storm-Petrels during spring and autumn were 6,400 and 713,900, respectively (95% CI: 4,500–10,000 in spring and 570,000–885,000 in autumn). Thus, abundance was ~112× greater in austral autumn than in spring.

Patterns Compared across Species

Annual Cycle of Abundant Migratory Species

Important in the determination of the annual cycle is information on molt, because molt does not overlap egg laying and overlaps little of the chick period, depending on whether the storm-petrel in question is sedentary or migratory (Ainley et al. 1976). However, for breeding-status comparisons of molt chronology, in which species were subdivided by month, only Leach's Storm-Petrels and *O. tethys* were abundant enough over several months to be included. Primary molt of both adult and subadult Leach's Storm-Petrels was stronger during boreal autumn than in spring (adults: $t = 10.96$, df = 143, $P < 0.0001$; Fig. 28A; subadults: $t = 5.87$, df = 187, $P < 0.001$; Fig. 28B), occurring primarily from October to December. By contrast, body molt of adult Leach's Storm-Petrels differed little between seasons ($t = 0.03$, df = 143, $P = 0.9$; Fig. 28C), whereas that of subadult Leach's Storm-Petrels was stronger in boreal spring than in autumn ($t = 6.30$, df = 187, $P < 0.001$; Fig. 28D). The latter timing is consistent with the detailed study of molt of this species at the Farallon Islands in central California (Ainley et al. 1976). Our findings suggest that the timing of events in the annual cycle of this species does not vary much by latitude, in spite of an immense spread in latitude of breeding sites (i.e., ~30°). The exception would be the winter breeding *O. l. cheimomnestes*. Consistent with the findings on molt, fat load of adult Leach's Storm-Petrels was heavier during boreal spring than in autumn ($t = 3.34$, df = 143, $P < 0.001$; Fig. 28E) (i.e., following completion of molt and before–during northward prebreeding migration), but that of subadults did not differ between seasons ($t = 0.07$, df = 187, $P = 0.9$; Fig. 28F).

Primary molt of both adult and subadult Wedge-rumped Storm-Petrels was stronger during austral autumn than in spring (adults: $t = 2.15$, df = 166, $P < 0.05$; Fig. 29A; subadults: $t = 2.49$, df = 87, $P < 0.01$; Fig. 29B). By contrast, body molt of adult and subadult Wedge-rumped Storm-Petrels differed little between seasons (t-tests, both $P > 0.1$; Fig. 29C, D). These results indicate broad overlap in molt with breeding in this form, in contrast to the highly migratory Leach's Storm-Petrel (above; Ainley et al. 1976). In similarity to the latter, fat load of adult Wedge-rumped Storm-Petrels was greater during austral spring than in autumn ($t = 3.81$, df = 166, $P < 0.001$; Fig. 29E), but that of subadults did not differ between seasons ($t = 0.64$, df = 87, $P = 0.5$; Fig. 29F).

Chronology of primary–body molt and fat load was similar among the sexes of Leach's and Wedge-rumped storm-petrels. To compare intensity of molt between the sexes, we conducted multiple regressions across the five storm-petrel species, in which we controlled for species differences, breeding status, and monthly differences. Results of these analyses indicated that, for a given month, breeding status, and species, females molted primaries more intensively than males ($t = 2.81$, df = 665, $P < 0.01$; Figs. 30A, B and 31A, B), though sex-related differences were not significant with

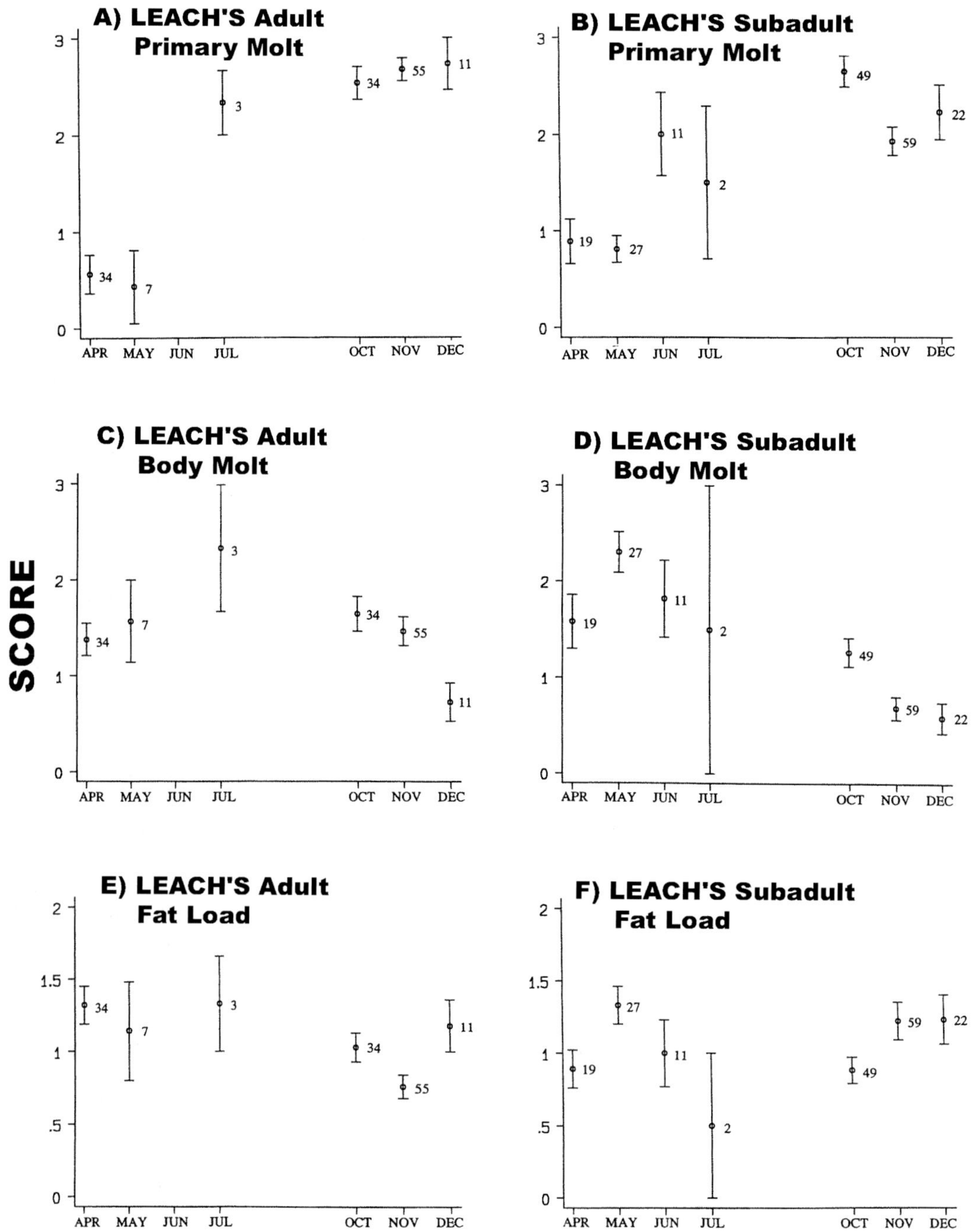

FIG. 28. Scores (means ± SE) for primary molt, body molt, and fat load (see text for information on scoring) in two age classes of light-rumped Leach's Storm-Petrel during spring and autumn, 1985–1995. Numbers adjacent to means denote number of birds examined.

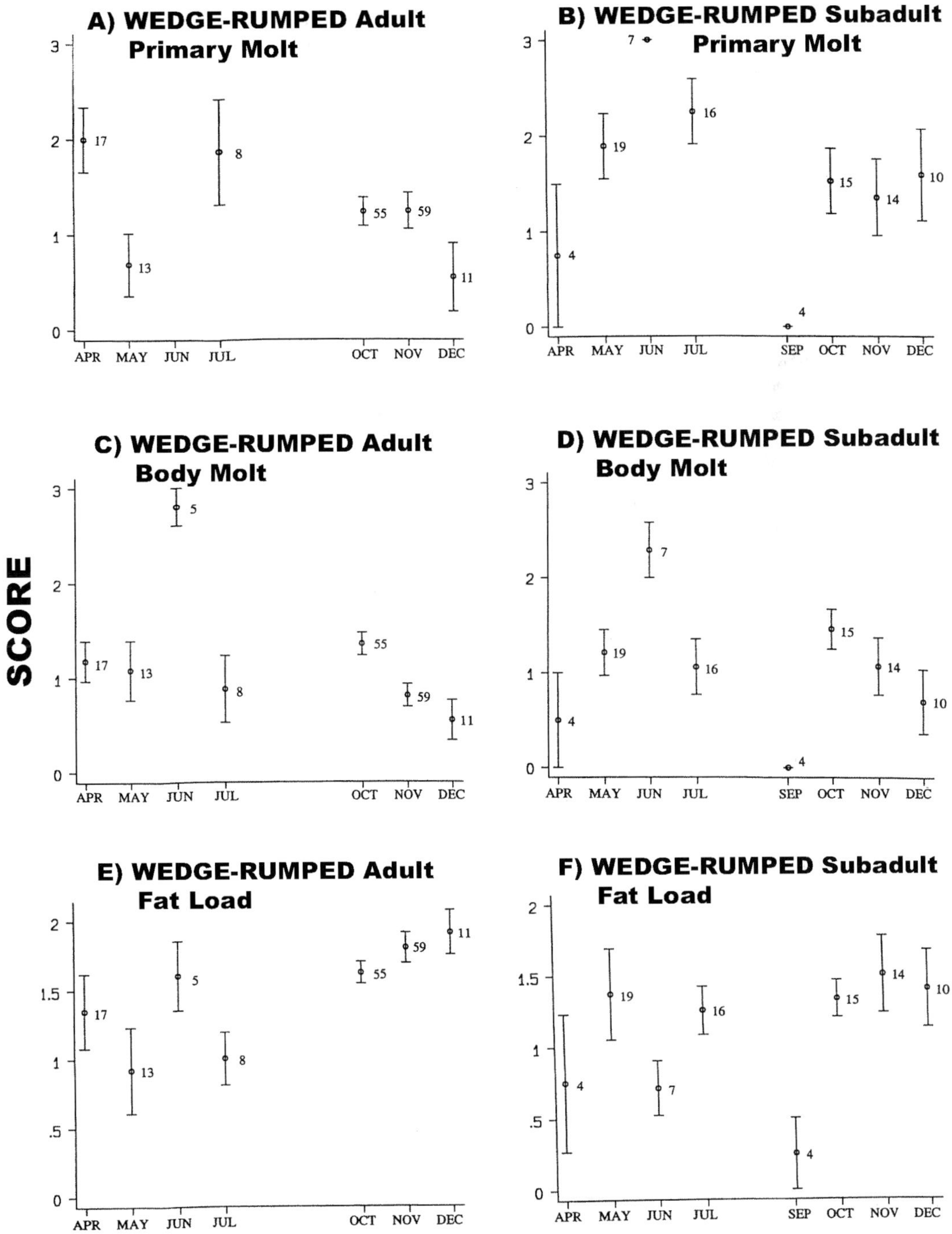

FIG. 29. Scores (means ± SE) for primary molt, body molt, and fat load (see text for information on scoring) in two age classes of Wedge-rumped Storm-Petrel (*O. t. tethys*) during spring and autumn, 1985–1995. Numbers adjacent to means denote number of birds examined.

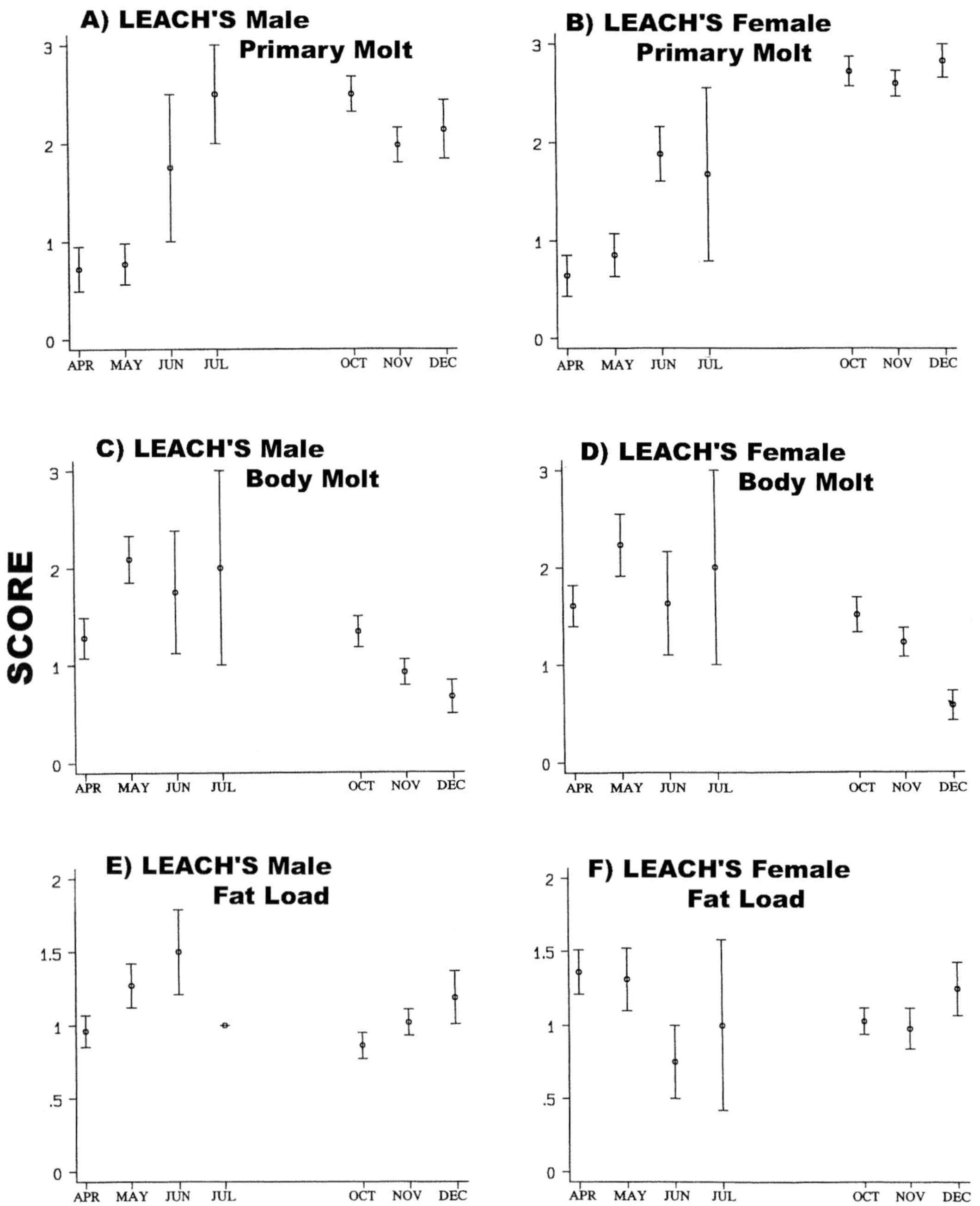

FIG. 30. Scores (means ± SE) for primary molt, body molt, and fat load (see text for information on scoring) in male and female light-rumped Leach's Storm-Petrel during spring and autumn, 1985–1995. Numbers adjacent to means denote number of birds examined.

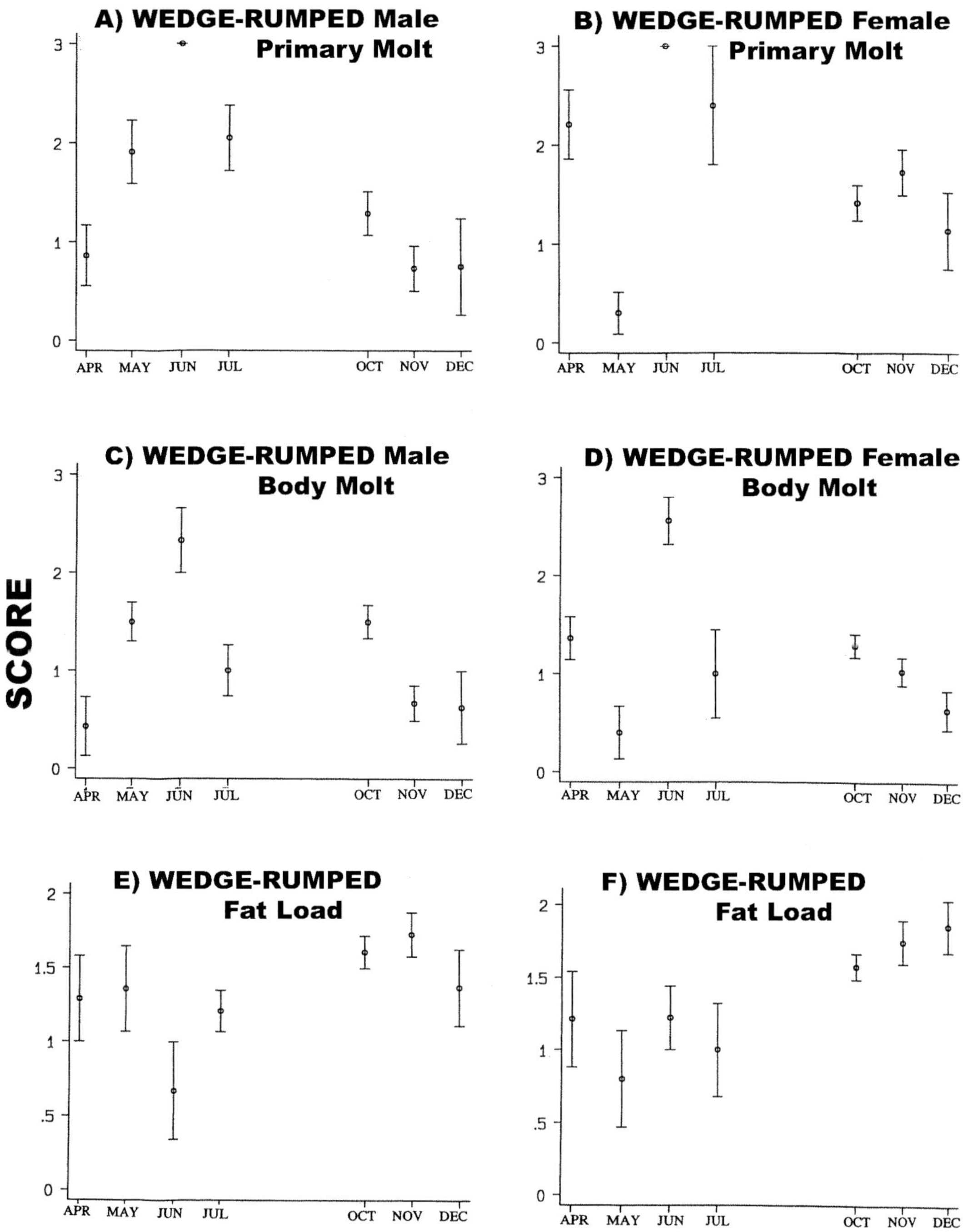

FIG. 31. Scores (means ± SE) for primary molt, body molt, and fat load (see text for information on scoring) in male and female Wedge-rumped Storm-Petrel (*O. t. tethys*) during spring and autumn, 1985–1995. Numbers adjacent to means denote number of birds examined.

respect to body molt ($t = 0.81$, $P = 0.4$; Figs. 30C, D and 31C, D) or fat load ($t = 0.08$, $P = 0.9$; Figs. 30E, F and 31E, F). These results reflect a more intense postbreeding molt among females, during boreal autumn for Leach's and during austral autumn for Wedge-rumped storm-petrels.

Habitat Use: Relationship to Environmental Gradients

For these analyses across taxa, assignment of data by season was hemisphere specific (Table 8 and Fig. 32). Thus, habitat use by northern-hemisphere breeders during boreal spring was compared with that of southern-hemisphere breeders during austral spring, and likewise for comparisons regarding autumn habitat use.

Sea-surface temperature.—During both spring and autumn, species occurring in association with the lowest and highest sea-surface temperature were Ashy and Least storm-petrels. Two other taxa associated with low and high temperatures, respectively, were Wilson's (*O. o. chilensis*) and White throated storm-petrels. Taxa most abundant in the Humboldt Current and Costa Rica Current boundary waters (White-vented, Markham's, Ringed, White-bellied, Band-rumped, White-faced, Black, and two races of Wedge-rumped storm-petrels) were usually associated with habitat of intermediate sea-surface temperatures.

Table 8. Ocean habitat characteristics (mean ± SD) of 15 taxa of storm-petrels in the eastern Pacific Ocean, 1980–1995. For each taxon, spring is represented by the upper row of values and autumn by the lower; seasons for northern- and southern-hemisphere breeders are boreal and austral, respectively. Values of *n* are the number of survey transects in which a given taxon was sighted; analyses for each taxon were weighted by storm-petrel density.

Taxon	Sea-surface temperature (°C)	Sea-surface salinity (ppt)	Thermocline depth (m)	Thermocline strength (Δ°C)	Wind speed (km h^{-1})	*n*
		Northern-hemisphere breeders				
Light-rumped Leach's	24.9 ± 5.2	34.21 ± 0.93	51 ± 27	6.0 ± 3.3	25 ± 14	1,578
	24.4 ± 3.5	34.32 ± 0.90	58 ± 33	4.9 ± 2.5	23 ± 10	2,326
Dark-rumped Leach's	19.4 ± 4.0	34.45 ± 0.79	45 ± 11	3.7 ± 1.4	40 ± 17	144
	27.2 ± 1.6	33.36 ± 0.54	31 ± 13	6.9 ± 1.7	20 ± 10	67
Ashy	13.1 ± 2.3	33.16 ± 0.60	22 ± 13	2.9 ± 1.6	14 ± 6	532
	14.0 ± 1.0	32.75 ± 0.64	51 ± 18	2.6 ± 1.2	14 ± 9	198
Black	24.4 ± 4.7	33.90 ± 0.65	31 ± 12	3.4 ± 1.9	18 ± 10	206
	23.1 ± 4.3	33.89 ± 1.13	15 ± 20	3.7 ± 2.3	14 ± 8	51
Least	28.0 ± 3.7	33.60 ± 0.30	25 ± 11	5.8 ± 0.9	8 ± 8	40
	28.3 ± 0.8	33.10 ± 0.81	38 ± 14	6.8 ± 1.3	14 ± 10	31
		Southern-hemisphere breeders				
Wedge-rumped	23.3 ± 2.6	34.72 ± 0.47	45 ± 30	3.8 ± 2.3	22 ± 8	471
(*O. t. tethys*)	26.5 ± 1.8	34.44 ± 1.23	30 ± 20	5.5 ± 2.6	20 ± 12	502
Wedge-rumped	22.0 ± 2.2	34.72 ± 0.63	23 ± 13	4.2 ± 2.2	21 ± 7	198
(*O. t. kelsalli*)	24.3 ± 2.8	34.67 ± 0.78	25 ± 1.9	4.5 ± 2.6	19 ± 8	659
Band-rumped	23.4 ± 2.7	34.33 ± 0.64	31 ± 11	4.7 ± 2.6	23 ± 9	82
	26.3 ± 1.7	34.25 ± 1.27	19 ± 15	5.4 ± 2.2	16 ± 12	355
Markham's	21.1 ± 1.4	35.39 ± 0.41	62 ± 37	3.3 ± 1.8	27 ± 11	199
	23.5 ± 2.7	34.85 ± 0.92	30 ± 19	4.9 ± 2.2	20 ± 9	456
Ringed	21.3 ± 1.4	35.39 ± 0.41	62 ± 37	3.3 ± 1.8	27 ± 11	87
	23.0 ± 1.7	35.10 ± 0.49	22 ± 10	3.4 ± 2.0	21 ± 7	112
White-vented	20.0 ± 1.0	35.04 ± 0.14	13 ± 14	3.5 ± 0.7	23 ± 6	94
	23.8 ± 3.1	34.92 ± 0.19	26 ± 14	2.3 ± 0.7	21 ± 7	164
Wilson's	16.6 ± 1.4	33.90 ± 0.86	21 ± 11	3.5 ± 0.7	28 ± 7	33
(*O. o. chilensis*)	15.2 ± 2.6	33.62 ± 1.01	25 ± 11	1.9 ± 0.5	15 ± 6	152
White-bellied	21.6 ± 2.6	35.30 ± 0.53	72 ± 27	3.5 ± 2.2	28 ± 8	51
(*F. g. segethi*)	21.3 ± 2.9	34.78 ± 0.58	70 ± 18	4.0 ± 2.3	20 ± 4	195
White-throated	25.5 ± 0.9	35.21 ± 0.32	84 ± 25	4.8 ± 1.6	30 ± 5	28
	27.7 ± 1.1	35.20 ± 0.58	70 ± 18	4.0 ± 2.3	25 ± 6	20
White-faced	21.4 ± 3.5	35.0 ± 0.63	30 ± 15	3.6 ± 1.4	19 ± 10	3
	24.2 ± 2.1	34.72 ± 1.08	48 ± 21	5.7 ± 2.3	27 ± 11	261

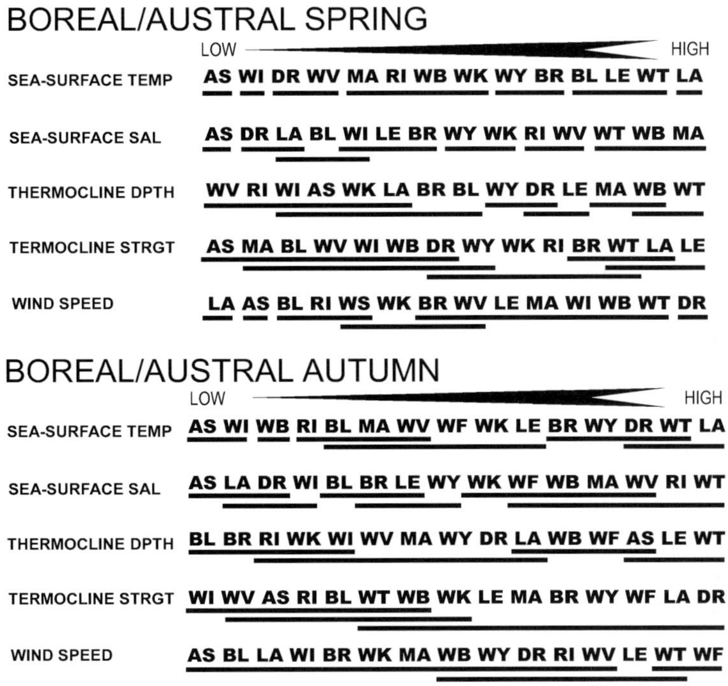

FIG. 32. Storm-petrels and habitat gradients: comparisons among 15 storm-petrel taxa with respect to their association with five habitat variables during spring and autumn, 1980–1995 (AS = Ashy Storm-Petrel, BL = Black, DR = dark-rumped Leach's, WV = White-vented, WY = Wedge-rumped [*T. t. tethys*], WK = Wedge-rumped [*T. t. kelsalli*], BR = Band-rumped, RI = Ringed, LA = Least, LE = light-rumped Leach's, MA = Markham's, WB = White-bellied [*F. g. segethi*], WF = White-faced (autumn only), WI = Wilson's [*O. o. chilensis*], WT = White-throated). Data were specific to the boreal or austral time-frame for northern- and southern-hemisphere breeders, respectively. Lines connecting taxa indicate insignificant differences (Sidak tests, $P > 0.05$). See Tables 8 and 9 for oceanographic data for each taxon.

Sea-surface salinity.—In both spring and autumn, the species most abundant in the California and Costa Rica currents (Ashy, dark-rumped Leach's, Least, and Black storm-petrels) were associated with lowest salinity, as were the Wilson's Storm-Petrels at the southern end of the Humboldt Current. On the other hand, species most abundant in the Humboldt Current (see above) were associated with higher salinity. The White-throated Storm-Petrel, most abundant in the SEC, was also associated with higher salinity.

Thermocline depth.—Although the White-throated Storm-Petrel was associated with the deepest thermoclines in both spring and autumn, there was considerable overlap among taxa in the association with thermocline depth; this was especially true during spring.

Thermocline depth and strength.—Although the White-throated Storm-Petrel was associated with the deepest thermoclines in both spring and autumn, there was considerable overlap among taxa in the association with thermocline depth and strength.

Wind speed.—During both spring and autumn, Ashy, Least, and Black storm-petrels were associated with the lowest wind speeds; White-throated, dark-rumped Leach's, and White-faced storm-petrels were associated with the highest wind speeds.

HABITAT USE: RELATIONSHIP TO COARSE-SCALE OCEANOGRAPHIC AND GEOGRAPHIC FEATURES

We used the first two PC axes of the PCA to compare overall use of environmental variables among the 15 storm-petrel taxa, because these axes explained 62% and 60% of the variation during spring and autumn, respectively (Table 9). In spring, the most important explanatory variables on the PC1 axis were positively loaded

TABLE 9. Principal component (PC) analyses: eigenvector (cumulative) proportions of variance explained by five habitat variables affecting at-sea distributions of 15 taxa of storm-petrels during spring and autumn, 1980–1995. Only oceanographic data from transect locations in which a storm-petrel of a given taxon was recorded were used (n = 3,746 for spring, n = 5,454 for autumn); analyses were weighted by storm-petrel density per transect. "Spring" is boreal spring for northern-hemisphere breeders and austral spring for southern-hemisphere breeders, and likewise for autumn (see Table 8 for taxa and sample sizes for each).

	Eigenvalue cumulative proportion			Eigenvector loadings			
				Spring		Autumn	
PC	Spring	Autumn	Variable	PC1	PC2	PC1	PC2
1	0.36	0.33	Sea-surface temperature	0.59	–0.42	0.56	–0.06
2	0.62	0.60	Sea-surface salinity	0.20	–0.08	–0.43	0.33
3	0.82	0.79	Thermocline depth	0.46	0.47	0.22	0.67
4	0.94	0.92	Thermocline strength	0.60	–0.18	0.64	–0.17
5	1.00	1.00	Wind speed	0.21	0.75	0.21	0.64

thermocline strength and sea-surface temperature. This is indicated by the elevated position of the Ashy Storm-Petrel on the negative side of that axis during spring (Fig. 33; see Fig. 32 for sea-surface temperature and thermocline strength for these taxa) and the elevated positions of Least, light-rumped Leach's, and White-throated storm-petrels on the positive side of the PC1 axis.

During spring, the most important variables on the PC2 axis were positively loaded wind speed and thermocline depth (Table 9). This is indicated by the elevated positions on the positive side of the PC2 axis of dark-rumped Leach's, White-throated, and White-bellied (*F. g. segethi*) storm-petrels, as well as the low positions on that axis of Least, Ashy, and Black storm-petrels (Fig. 33; see Fig. 32 for thermocline depth and wind speed for these taxa).

Overall, overlap was minimal in the habitat use among storm-petrel taxa during spring (Fig. 33). The only taxa that did not differ significantly from each other were three Humboldt Current endemics: Ringed and White-vented storm-petrels, and the *O. t. kelsalli* race of Wedge-rumped Storm-Petrel.

During autumn, just as during spring, the most important variables on the PC1 axis were positively loaded thermocline strength and sea-surface temperature (Table 9). This is indicated by the elevated position of Least and dark-rumped Leach's storm-petrels on the positive side of the PC1 axis during autumn (Fig. 34; see Fig. 32 for sea-surface temperature and thermocline strength for these taxa) and the elevated positions of the *O. o. chilensis* race of Wilson's and Ashy storm-petrels on the negative side of that axis.

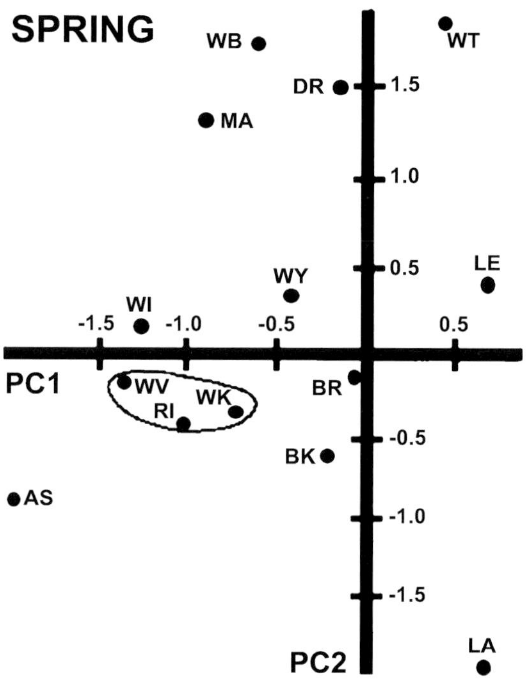

FIG. 33. Results of PCA comparing associations of 14 storm-petrel taxa with five environmental variables during spring, 1980–1995. Data were specific to the boreal and austral seasons for northern- and southern-hemisphere breeders, respectively. See Figure 32 for definitions of species' acronyms, and Table 8 for sample sizes. White-faced Storm-Petrel is excluded from this analysis, because of low sample size in spring.

The most important variables on the PC2 axis during autumn, as with spring (above), were positively loaded wind speed and thermocline depth (Table 9). This was indicated by the elevated position of White-throated, light-rumped Leach's, and White-faced storm-petrels on the positive side of the PC2 axis; by the high position on the negative side of that axis of dark-rumped Leach's, Least, Black, and Band-rumped storm-petrels; and by the low positions of Markham's and Wedge-rumped storm-petrels (Fig. 34; see Fig. 32 for thermocline depth and wind speed for those taxa).

During autumn, overlap in habitat use was similar among two species-groups of Humboldt Current endemics: White-bellied (*F. g. segethi* race), White-vented, and Ringed storm-petrels in one group; and Markham's Storm-Petrel and the *O. t. kelsalli* race of Wedge-rumped Storm-Petrel in the other (Fig. 34). Habitat use by light-rumped Leach's and White-faced storm-petrels also did not differ significantly during autumn. The White-faced Storm-Petrel occurred primarily in the outer Humboldt Current but also in the SEC, where light-rumped Leach's Storm-Petrel is most abundant during autumn (see species accounts).

For northern-hemisphere breeders, habitat use did not differ between boreal spring and autumn among light-rumped Leach's, Black, and Least storm-petrels, but differed significantly between seasons among dark-rumped Leach's and Ashy storm-petrels (Sidak tests, all $P < 0.025$). Habitat use among the 10 taxa of southern-hemisphere breeders did not differ seasonally among Markham's, White-throated, and the *O. t. kelsalli* race of Wedge-rumped storm-petrels, but differed significantly between

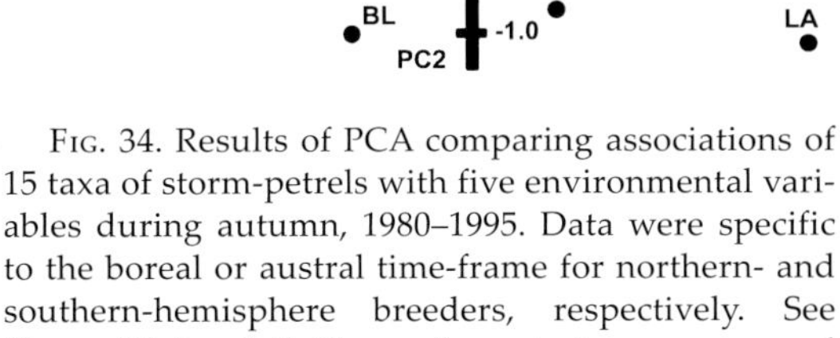

Fig. 34. Results of PCA comparing associations of 15 taxa of storm-petrels with five environmental variables during autumn, 1980–1995. Data were specific to the boreal or austral time-frame for northern- and southern-hemisphere breeders, respectively. See Figure 32 for definitions of species' acronyms, and Table 8 for sample sizes.

seasons among each of the others (Sidak tests, all $P < 0.025$).

Discussion

Below, we will discuss the factors that led to the diversification of storm-petrels in the eastern Pacific Ocean and those involved in whether various species can coexist, and at what scale. First, though, we will discuss certain preliminary subjects: overall taxon abundance, overlap in annual cycles, surplus or "floating" populations, population movements from breeding to nonbreeding areas (therefore, changes in overlap), foraging patterns, and, finally, ocean habitat differences.

Abundance

The present study included all storm-petrel species that occur in the eastern Pacific Ocean, with the exception of three species whose ranges we barely intersected (see above) and, of course, the Guadalupe Storm-Petrel, which was extirpated in historical times. Although we provide the first abundance estimates ever done for 18 taxa, it is important to note that we had complete survey coverage of the ranges of only 10: Ringed, White-vented, Markham's, the *O. t. segethi* race of White-bellied, both races of Wedge-rumped, the Galápagos race of Band-rumped (*O. c. bangsi*), and races of dark-rumped Leach's storm-petrels (*O. l. socorroensis*, *O. l. cheimomnestes*, and *O. l. chapmani*). During boreal autumn, we also had nearly complete coverage of the range of Black and Least storm-petrels, because nearly all vacate the Gulf of California after the breeding season (Howell and Webb 1995, Ainley and Everett 2001). We also had nearly complete coverage of the pelagic range of the White-throated Storm-Petrel in the Pacific.

Our at-sea abundance estimates of light-rumped Leach's, White-bellied (*F. g. grallaria*), and White-faced storm-petrels are within the range of those expected from colony-based estimates (see above); those of Black and Least storm-petrels are lower, whereas those of all other taxa are greater, particularly among species nesting on the Galápagos Islands and at sites along the coast of South America. Lower abundance among Black and Least storm-petrels may reflect previous population overestimation but could also be attributable to our lack of surveys in the Gulf of California during spring (i.e., incomplete coverage of those species' ranges when breeding). For the dark-rumped Leach's, Band-rumped, both races of Wedge-rumped, White-vented, Ringed, Markham's, Chilean Wilson's (*O. o. chilensis*), and Chilean White-bellied (*F. g. segethi*) storm-petrels, our considerably higher at-sea-based estimates, as compared with colony-based estimates, likely reflect previous underestimation at colonies owing to the difficulty of finding storm-petrel nests and the likelihood of "floating" populations (see below), combined with the inaccessibility of the desert areas that occur within 200 km of the coast where the South American species nest.

The populations of all but seven of the storm-petrel taxa residing in the eastern Pacific are very large, numbering from low hundreds of thousands to millions. The light-rumped Leach's Storm-Petrel (*O. l. leucorhoa*) is, by far, the most abundant storm-petrel in the eastern Pacific. Among the seven forms with low abundance, White-throated, a race of White-bellied (*F. g. grallaria*), a race of Wilson's (*O. o. oceanicus*), and White-faced occupy large areas in very low densities primarily on the western side of the eastern Pacific; the Ashy Storm-Petrel is confined to a small region in the California Current; and the pelagic ranges of *F. g. titan* and *O. c. cryptoleucura* remain largely unknown. All except Wilson's are represented by only a few breeding colonies.

Previous estimates for population size of some of these less-abundant forms were rough. Yet our estimate of the pelagic population of *F. g. grallaria* (2,866 birds; 95% CI: 2,200–5,700), which includes the nonbreeding component as well as breeders, is consistent with estimates from known breeding sites at Kermadec Island and islets near Lord Howe Island, totaling roughly 4,000 breeders (Fullagar et al. 1974, Robertson and Bell 1984, Marchant and Higgins 1990). Apparently, *F. g. grallaria* has been extirpated from Lord Howe Island itself, possibly as a result of predation by feral mammals. The absence in our study of sightings of melanistic *F. g. grallaria*, including the 17 birds we collected, makes assigning the group of *F. g. grallaria* using the west side of the eastern Pacific to a given breeding population difficult, given that both the Kermadec and, particularly, Lord Howe populations include large proportions of melanistic birds.

Our estimates for White-throated (52,300 birds; 95% CI: 35,500–87,900) and Ashy (7,290 birds; 95% CI: 4,500–9,000) storm-petrels are somewhat larger than those made at known breeding locations (Garnett 1984, Sydeman et al. 1998). For some of these less-abundant forms, including the Ashy Storm-Petrel, we had incomplete range coverage, which would negatively affect our estimates of the total population. Nevertheless, this study confirms the need for continued or increased measures of protection for these birds. The most endangered population in our study area appears to be *F. g. titan*, of which we observed one bird during the entire study, despite the fact that this form is readily identifiable at sea by its size, which is very large for a White-bellied Storm-Petrel (Appendix 2). Indeed, *F. g. titan* may represent a distinct species. The remnant population of this bird, which nests at Rapa Island (Holyoak and Thibault 1984), requires strict protection if extinction is to be avoided.

Annual Cycle and Breeding Populations

In the following, we assume that the pattern of molt for mobile and sedentary species follows the model described in Ainley et al. (1976: for Leach's, a highly mobile form, and Ashy, a sedentary form; see also review in Warham 1990, 1996). In the migratory forms, body molt begins just as chicks are fledging, and wing molt does not overlap migration; in the sedentary forms, body molt extensively overlaps chick provisioning, and wing molt overlaps the end of the current nest-site occupation and the beginning of the next. We are aware of just one study that has compared molt patterns among sexes and age classes of a Procellariiform species, that of Weimerskirch (1991) on the Wandering Albatross (*Diomedea exulans*), a study that had mixed results (reviewed in Warham 1996).

The group of storm-petrel species that we collected and examined for breeding-status-related molt patterns spends the majority of its sea time in the equatorial Pacific, though none of them can be considered true trans-equatorial migrants. Our samples of the two most abundant species, Leach's Storm-Petrel and *O. t. tethys*, allowed us to examine molt on a month-by-month basis. Molt chronology of the Leach's Storm-Petrel was consistent with a highly migratory form, and that of the Wedge-rumped Storm-Petrel was consistent with a species that merely disperses in the vicinity of the colony during the nonbreeding season. Otherwise, in general, molt of the eastern Pacific storm-petrels occurred mostly in the postbreeding period, though both of the species just mentioned apparently have a phase of heavy prenuptial body molt. In contrast to adults, the body molt of subadult Leach's apparently begins earlier in the year, on average, than that of breeders, though we know of no other studies showing similar age-related patterns in petrels. In addition, female Leach's and Wedge-rumped storm-petrels, though the overall timing of their molt was similar to that of males, demonstrated a more synchronous primary molt (but not body molt) than males. Thus, the indication was that females probably finish molt earlier than males, which for both species would mean that molt of males may tend to continue into the boreal or austral winter preceding the onset of colony occupation.

Our information based on molt and other factors indicated consistency with the known breeding seasons for most eastern Pacific Ocean taxa (see species accounts). However, we also presented the first definitive evidence for the likelihood of a spring–summer breeding season in the Ringed Storm-Petrel.

Although we would expect populations to be represented at sea by higher numbers during the nonbreeding season (when breeders, nonbreeders, and recent fledglings should be at sea) than during the breeding season (if known), this was frequently not the case (Table 10). For example, three taxa (Markham's, Ringed, and the Wedge-rumped storm-petrel race *O. t. kelsalli*) of the five for which we had complete coverage of the pelagic range had highest abundance during the breeding period. These exceptions can result only from three factors: (1) we did not have complete coverage of the taxon's range during both seasons; (2) the breeding season is less well known than thought; and (3) significant "floating" populations exist among breeders, all of which occur in the vicinity of breeding sites during the breeding season (in the nonbreeding season, they disperse to areas insufficiently sampled by us). Inaccurate identification is not likely to be a factor. In the case of Markham's and Black storm-petrels, the possibility that misidentification was involved was unlikely, because of the low abundance of Black Storm-Petrels within the range of Markham's Storm-Petrels and the fact that the two can, with practice, easily be distinguished (see above). Regarding races of Wedge-rumped Storm-Petrel, which cannot be distinguished at sea, we do not suspect misidentification as a factor, because we did not depend on "identification"

TABLE 10. Completeness of survey coverage, breeding season, season of maximum abundance, seasonal difference in number of birds, and seasonal difference as percentage of seasonal maximum for 15 taxa of storm-petrels in the eastern Pacific, 1980–1995. (A Autumn = austral autumn, B Spring = boreal spring, etc. See Table 6 for seasonal population estimates.)

	Complete coverage?	Breeding season	Season maximum abundance	Seasonal difference	Difference (%)
		Northern-hemisphere breeders			
Light-rumped Leach's	No	B Spring	B Autumn	2,063,400	34
Dark-rumped Leach's	Yes	B Spring	B Autumn	518,600	43
Black	No	B Spring	B Autumn	120,300	21
Ashy	No	B Spring	B Spring	3,080	42
Least	No	B Spring	B Autumn	113,800	37
		Southern-hemisphere breeders			
Band-rumped	Yes	Bi-seasonal	A Autumn	353,000	74
Wedge-rumped					
O. t. tethys	Yes	A Autumn	A Spring	508,700	45
O. t. kelsalli	Yes	A Autumn	A Autumn	193,100	13
Markham's	Yes	A Autumn	A Autumn	293,500	27
Ringed	Yes	Prolonged	A Spring	374,700	37
White-vented	Yes	Prolonged	A Autumn	462,300	54
Wilson's					
O. o. chilensis	No	A Autumn	A Spring	460,600	53
White-bellied					
F. g. segethi	Yes	A Spring	A Autumn	327,900	74
White-throated	No	Prolonged	A Spring	31,475	60
White-faced	No	A Spring	A Autumn	707,500	99

during surveys to differentiate populations. Instead, we used identification of collected birds to determine range limits of each taxon.

We suspect that some cases in which breeding-period populations were larger than nonbreeding-period populations are explained, at least in part, by incomplete knowledge of the extent of the breeding season. The seasonal difference in population estimates was relatively small for Wedge-rumped *O. t. kelsalli* (13%), but larger (27%) for Markham's and Ringed storm-petrels. Information on the breeding period of *O. t. kelsalli* has been recorded from several nesting locations on the coast of Peru (Murphy 1936); that of Ringed and Markham's storm-petrels has been described only very recently. In the case of the latter, reports were based on the discovery in 1992 of colonies in one discrete area (Paracas Peninsula; Jahncke 1993). Yet, judging from the at-sea abundance of this species between northern Peru and northern Chile during spring and autumn (Fig. 19), its breeding localities are likely spread more extensively than just that one location. Similarly, we assume an austral spring–summer breeding season for the Ringed Storm-Petrel, based on evidence of the timing of egg laying in a sample of three birds. Therefore, we believe that the breeding seasons of *O. t. kelsalli* and Ringed and Markham's storm-petrels may be more protracted than suspected, or even biseasonal (i.e., similar to those of Band-rumped, Ringed, and White-vented storm-petrels [details below], the other species presumably nesting in northern South America).

In regard to the White-vented Storm-Petrel, most evidence indicates that its egg-laying to fledging period extends from May (indicated by specimens with enlarged gonads and an egg; Murphy 1936) to as late as November (indicated by a bird found incubating an egg in August; Schlatter and Marin 1983). Thus, the breeding season appears to be prolonged, lasting from austral autumn through spring, though the higher at-sea abundance we observed in austral autumn favors a breeding season centered in austral spring and summer. The breeding season of Ringed Storm-Petrels may also be prolonged. Birds collected in May through August were in nonbreeding condition and were undergoing what should have been their postbreeding primary molt (Murphy 1936, Crossin 1974). However, the recording of grounded fledglings in coastal Peru and Chile in June and July indicates an austral autumn-to-winter breeding season extending from March to July (Brooke 2004). In late November off Peru, however, we collected a female Ringed Storm-Petrel that contained an egg nearly completely formed, which suggests that some birds begin breeding in austral summer.

We can only suggest that further study is needed to better document the breeding seasons for each of the four species just discussed. This should include full coverage of the breeding range, which, judging from coastal distributions, appears to extend along the South American coast from about 7°S to 17°S in *O. t. kelsalli* and White-vented Storm-Petrels, and from about 8°S to 25°S in Ringed and Markham's storm-petrels.

The White-throated Storm-Petrel's breeding season is also believed to be nonseasonal, whereas the Band-rumped Storm-Petrel race *O. c. bangsi* breeds biseasonally (Murphy 1936, Harris 1969, Brooke 2004). However, our abundance estimates of Band-rumped and White-throated storm-petrels were highest in the austral autumn and spring, respectively, which leads us to speculate that the primary breeding period is likely to be during austral spring for the former and austral autumn for the latter.

There were two other taxa, Ashy Storm-Petrel and the Wilson's Storm-Petrel race *O. o. chilensis*, for which breeding-season estimates were higher than nonbreeding-season estimates, though we had near-complete survey coverage only of the Ashy Storm-Petrel (at least of the central California population). One concentration of breeding in that species is at the Farallon Islands, off central California; the other is on the Channel Islands (CI), off southern California, 1,000 km south of the Farallon Islands. We suspect that the higher estimate during the breeding season for the more northern, Farallon breeding area is likely attributable to (1) birds from CI moving north during spring, as is the case with other species from those waters (e.g., Black Storm-Petrel, Brown Pelican [*Pelecanus occidentalis*], Heermann's Gull [*Larus heermanni*], Xantus's Murrelet [*Synthliboramphus hypoleucus*]; Howell and Webb 1995); (2) part of the Farallon population moving south during autumn; or (3) a sizeable "floating" population.

Regarding *O. o. chilensis*, although this taxon is suspected of breeding on the Falkland

Islands, it has seldom been recorded anywhere else in the Atlantic Ocean (Murphy 1936). Therefore, we suspect that the larger numbers on the Pacific coast of South America during the breeding season reflect (1) extensive movements of nonbreeders to the western coast of South America during the austral spring breeding season, in combination with movement by these birds back to the southern tip of the continent during austral autumn; and (2) lack of postbreeding dispersal by a large number of the breeding population from the breeding areas on Tierra del Fuego and Cape Horn (areas we did not survey during the present study).

Finally, floating populations (i.e., portions of populations composed of adult birds that are capable of breeding but do not, because they lack a breeding site; e.g., Manuwal 1974) are likely among storm-petrels of the eastern Pacific, especially because most rely on natural cavities for nesting; any taxon in which we found breeding populations higher than nonbreeding ones is a candidate for having such a population structure. The importance of and competition for various breeding sites is exhibited by their year-round use in some species. Nest-site competition could be fierce. Distinct summer- and winter-breeding populations occur on both Guadalupe Island (*O. l. socorroensis, O. l. cheimomnestes*) and the Galápagos Islands (*O. c. bangsi*, two populations), and some evidence exists that this could be so in the Humboldt Current as well, as indicated by *O. hornbyi* and, perhaps, other taxa discussed above. In further support of the hypothesis that nesting space is limited, thus encouraging the existence of floating populations, are the prolonged nesting seasons of several forms both in the California and Humboldt current systems (e.g., Ashy, Markham's, and White-vented storm-petrels maintain a territory well before or after the breeding season to preclude losing it to a competitor). In the Gulf of California, nesting areas are in such short supply that Least and Black storm-petrels nest in appropriately sized arms of cavities that have the same entrance, sharing the site with other birds and mammals as well (Ainley 2005).

Movement Patterns

Three types of movement patterns have been observed among seabirds (Spear 2001). These include "true migration," in which all members of a population move from a breeding area to a wintering area disjunct from the former; "partial migration," in which some members of a population migrate and others do not; and "dispersal," in which individuals move various distances after the breeding season, such that they occur at all distances within a given radius of the breeding site.

Eastern Pacific storm-petrels are primarily dispersers, particularly the Humboldt and California current endemics (Wedge-rumped [*O. t. kelsalli*], Markham's, Band-rumped, Ringed, Elliot's, White-bellied [*F. g. segethi*], Wilson's [*O. o. chilensis*], dark-rumped Leach's, Ashy, and White-throated storm-petrels). Partial migrants include the light-rumped Leach's, Fork-tailed, Black, Least, Wilson's (*O. o. oceanicus*), and probably the White-faced and White-bellied (*F. g. grallaria*) storm-petrels. The only really long-distance migrants among the forms that we studied are the Leach's Storm-Petrel (*O. l. leucorhoa*), and the Antarctic-breeding Wilson's Storm-Petrel (*O. o. oceanicus*). The former nests from northern California north to and among the Aleutian Islands and winters in the equatorial region; the latter nests in the low-latitude Antarctic, and, to some degree, winters in the subtropical Pacific (and Atlantic). Other races of the Leach's, all of which nest from central California south, are more similar in their movements to partial migrants. The same is true for *O. o. chilensis*. The molt pattern of *O. t. tethys* is similar to that of a species that moves extensively, perhaps as a partial migrant or a very wide-ranging disperser.

Behavior at Sea

We report here some of the first information on certain aspects of the behavior of storm-petrels at sea, other than the actual method of prey capture (see Ainley 1977). After feeding, seabirds generally become inactive for a period, forming groups of resting birds sitting on the ocean surface (D. G. Ainley and L. B. Spear pers. obs.). On average, eastern Pacific Ocean hydrobatid taxa used ~45% of the daylight hours resting on the water and feeding, as opposed to flying in transit. This proportion of resting is greater than that of larger procellariiforms, such as the diomedeids and procellariids we studied in the eastern Pacific Ocean, but not than that of the tropicbirds (Phaethontidae; Spear et al. 1995a, b, 2003; Spear and Ainley 2005b). Three species of albatross wintering in the Humboldt Current used ~30% of the day resting–feeding, as did four species of procellariids in the equatorial Pacific. Tropicbirds used ≥60% of their daily time resting and feeding (i.e., not flying in transit to or from nesting colonies or in search of food).

These differences are likely related to flight energetics and feeding methods. Of the four families, flight efficiency is poorest among tropicbirds, which are large and have high wing loading and, therefore, use flapping flight strictly (Pennycuick 1989, Spear and Ainley 1997b). However, most procellariiforms use the more efficient gliding or flap-gliding flight and make more efficient use of the wind (Spear and Ainley 1997b). Among petrels, flight efficiency increases with body size; therefore, we would expect that albatrosses would use the least amount of time resting and feeding, that procellariids would be intermediate, and that hydrobatids would use these behaviors most often.

Food-search methods also differ considerably among the three families of procellariiforms. Using primarily gliding flight, the albatrosses and procellariids search widely over ocean areas in which food is very patchy; but having found food, they feed extensively and have a relatively nondiverse diet (reviewed in Spear et al. 2007). By contrast, storm-petrels, using primarily flapping flight, also widely search ocean areas for patchy prey but have a more diverse diet. However, given their much slower flight speed, the area of their search is far smaller than that of the larger petrels. Hydrobatids are also the smallest procellariiforms and, thus, require an appreciable energy intake per unit of body mass to maintain homeothermy (Nagy 1987). The energetics of their small size is likely the factor that precludes storm-petrels from the very highest latitudes.

Northern-hemisphere breeders, including light- and dark-rumped Leach's, Black, and Least (but not Ashy) storm-petrels, had a higher frequency of feeding–resting (vs. flying in transit) during boreal spring than in autumn. This may be related to the fact that these species all breed during spring–summer and, therefore, are feeding young at that time. By contrast, 8 of the 10 southern-hemisphere breeders, including Wedge-rumped (both races), Markham's, Ringed, White-vented, Wilson's (*O. o. chilensis*), White-bellied (*F. g. segethi*), and White-faced storm-petrels, had higher incidence of feeding–resting during austral autumn than in spring. However, unlike the northern-hemisphere species, feeding incidence was not consistent with timing of breeding, given that three of the southern-hemisphere species listed above breed in austral autumn, three breed in spring, and two cannot be assigned to season. Exceptions to the seasonal pattern of the eight southern-hemisphere species were the Band-rumped Storm-Petrel, a biseasonal breeder whose feeding incidence differed little between seasons, and the White-throated Storm-Petrel, a nonseasonal breeder that had higher feeding incidence during austral spring.

Environmental Factors Affecting at-Sea Distributions

Each storm-petrel taxon exhibited characteristic current-system and ocean-habitat preferences and sorted along environmental gradients. Six taxa were strictly pelagic (deep ocean areas off the shelf), including light-rumped Leach's, Wedge-rumped (*O. t. tethys*), Wilson's (*O. o. oceanicus*), White-bellied (*F. g. grallaria* and *F. g. segethi*), and White-throated storm-petrels; White-faced Storm-Petrel was primarily pelagic. Six other taxa occurred in highest abundance over the continental shelf and, hence, would be termed neritic, including Least, Ringed, White-vented, Markham's, and Wedge-rumped (*O. t. kelsalli*) storm-petrels. Three taxa were most abundant over the continental slope: dark-rumped Leach's, Black, and Band-rumped (*O. c. bangsi*) storm-petrels. The few remaining taxa had broader habitat preferences. Ashy Storm-Petrels occurred in highest abundance over slope and pelagic waters in equal proportions. Taxa preferring the continental-shelf and slope habitats were residents of the inshore current systems—the California, Costa Rica, and Humboldt currents—whereas pelagic taxa occurred primarily in the three equatorial current systems, but primarily the SEC.

Not surprisingly, specific use of different depth zones and current systems by the eastern Pacific Ocean hydrobatids was consistent with their associations with various oceanographic parameters. In the regressions performed for each taxon, by far the most important oceanographic variable affecting their distributions was sea-surface temperature. These results were also supported by results of the PCA performed to identify important oceanographic variables among storm-petrels as a group. That analysis indicated that during both spring and autumn (boreal time-frame applied to northern-hemisphere breeders, and austral time-frame applied to those of the southern hemisphere), the most important variables were sea-surface temperature, thermocline structure of the water column, and wind speed. These variables are generally proxies for ocean productivity, gradients of which are important in the study area (Ballance et al. 1997, Spear et al. 2001, Vilchis et al. 2006).

Yet, except for the Humboldt Current endemics, which include White-vented, Ringed, Markham's, Wedge-rumped (Galápagos *O. t. kelsalli*), and White-bellied (*F. g. segethi*) storm-petrels, all taxa except light-rumped Leach's and White-faced storm-petrels were distinct from one another in their associations with different oceanographic parameters. This was particularly true of the Least, White-throated, Ashy, Black, dark-rumped Leach's, and Wilson's (*O. o. chilensis*) storm-petrels. For Black, dark-rumped Leach's, and Least storm-petrels, their distinctness was partially related to an association with warm waters of the Costa Rica Current, especially in the Gulf of Panama, where all, but particularly Least Storm-Petrels, are very abundant (Spear and Ainley 1999).The White-throated Storm-Petrel was also associated with warmer waters with a very deep thermocline. This species was one of the highly pelagic taxa and was most abundant on the western side of the SEC, where that current is warmest and has a deep thermocline (Wyrtki 1966). By contrast, the association of Ashy and Wilson's (*O. o. chilensis*) storm-petrels with cooler water, compared with the other taxa, was consistent with their predominant use of upwelling regions in the California and Humboldt currents, respectively. Least and Ashy storm-petrels were also associated with low wind speed, compared with the White-throated and dark-rumped Leach's storm-petrels, which were associated with higher wind speed.

Why Is Storm-petrel Diversity High in the Eastern Pacific Region?

Judging from the regional diversity or number of species (i.e., gamma diversity), storm-petrels likely arose—or at least the group experienced its greatest phylogenetic radiation—in the Pacific Ocean, and perhaps in the eastern Pacific. There are 19 species of storm-petrel breeding in the Pacific (all but a few in the eastern Pacific), compared with just 8 in the Atlantic, 3 in the Indian, and 3 in the Southern oceans (maps in Harrison 1983). Except for just one species in the Atlantic (*Hydrobates pelagicus*), these other oceans share their species with the Pacific. Although the Pacific is far larger (and older) than these other oceans, which may affect the number of species present, it is not 2.5–6 times larger. Six species (plus one recently extirpated, Guadalupe Storm-Petrel) are found nowhere else but in the upwelling current systems of the eastern Pacific; eight distinct taxa nest on islands in the California Current, and four distinct taxa nest on islands in the Humboldt Current.

If the extent of speciation, as in tropical rainforests, can be attributed, in part, to the age and stability of a habitat, then the greater diversity of storm-petrels should reflect the long-time existence of the oceanographic systems in the eastern Pacific (Pianka 1966, Brown 1981). The current systems and water masses of the eastern Pacific, as presently structured, are indeed old, having arisen in the late Miocene, 10–8 mya (Warheit 1992). Storm-petrels as species in the eastern Pacific are about that old. In fact, storm-petrel fossils of the genus *Oceanodroma* appear in coincidence with the appearance of the eastern Pacific upwelling systems, a period that also saw a general increase in seabird species diversity in the region (Warheit 1992). The genus *Oceanites* appears to be more recent, judging from the younger age of the few fossils thus far found (work of S. Olson, summarized in Warheit 2002), and it has not radiated as much as is evident in *Oceanodroma*. Fossils of the other extant storm-petrel genera are not known.

The physical attributes of the eastern Pacific, and especially the steep gradients in a number of environmental variables in accordance with the age and stability of its current systems, are also highly conducive to strong species radiation. This idea of geographic variation in response to environmental gradients, ultimately leading to speciation, was initially synthesized by Mayr (1964) and has been "proven" in many more recent studies of birds, amphibians, and other vertebrates and their genetics (e.g., Garcia-Paris et al. 2000, Genovart et al. 2003, Rundle and Shluter 2004, Wikelski 2005). These patterns are apparent among the eastern Pacific Ocean storm-petrels, though the complexity of the situation is increased by virtue of many storm-petrels in this region nesting allopatrically but mixing strongly at sea during nesting or in wintering areas. A similar sort of situation was described by Grant (1986) among certain of the Galápagos finches (see also Cody 1974).

Another attribute of the eastern Pacific that would affect species radiation is that the availability of nesting habitat is somewhat clumped across the different ocean regimes, with wide stretches of ocean between clumps. These nesting clumps include the coastal islands in the California Current, Gulf of California islands, oceanic Guadalupe Island in northern subtropical waters, oceanic Galápagos Islands in southern subtropical waters, and the Peru guano islands in the Humboldt Current. Also included are "habitat islands" in the Peruvian–Chilean desert. Scattered within this clumping of breeding habitat are a smattering of sites where a number of distinct storm-petrel taxa nest (e.g., Farallon-Channel Islands [one distinct taxon], Guadalupe Island [among distinct taxa, two extant plus one recently extirpated], and the Galápagos Islands [three distinct taxa]). Coastal islands, where other storm-petrels now nest, would have been bridged to the continent during the repeated ice ages that occurred since the Miocene, thus precluding storm-petrel breeding. However, a number of sea mounts occur in the vicinity of these locations, and several would have been exposed during the low sea levels of the ice ages, thus providing alternative breeding habitat.

Given the low incidence of large-scale movement among several eastern Pacific storm-petrels, as demonstrated in this report, such clumping of breeding habitat would have further encouraged species isolation and radiation because of the intense habitat gradients in the surrounding ocean. Indeed, ocean productivity gradients isolate these island clumps. Owing to the large-scale physiography of the eastern Pacific, the various nesting centers are distributed within an ocean area unparalleled in its oceanographic heterogeneity, including two highly productive eastern boundary currents, the convergence of two large ocean gyres with countercurrents in between, and a series of distinct coastal currents (e.g., Costa Rica Current; Longhurst and Pauly 1987; see Fig. 1). A highly productive environment, as demonstrated in terrestrial situations, allows avian species to partition the habitat more finely than they would be able to otherwise (Pianka 1966, Brown and Maurer 1987).

Our analysis indicated a further clear partitioning of the ocean among the various 26 forms of storm-petrels of the eastern Pacific. We found especially that, among the endemic storm-petrel avifauna in the Humboldt and California Current regions, various species were sorted along mesoscale depth gradients (or inshore–offshore gradients), which certainly relate to ocean productivity—from the high productivity of waters over continental shelves, to the continental slope, and finally to the low-productivity waters overlying the deep ocean. In fact, during the prehistoric period, when eastern Pacific upwelling currents appeared and when storm-petrels radiated as species, the eastern Pacific upwelling currents were much narrower than they are now, thus intensifying these gradients (Warheit 1992). During interglacial periods (e.g., present time), this sorting and any competition for habitat could well be relaxed. With the absence of continental shelves with low sea levels during glacial maxima, however, storm-petrel habitat along the Pacific Ocean's eastern boundary would have been severely squeezed, thus increasing competition for foraging space. Therefore, from a shelf to slope to pelagic perspective (i.e., high to low ocean productivity), the following gradients in occurrence at the species level are evident as one travels from north to south in the study area: Ashy–Leach's, Black–Ashy–Leach's, Black–Least–Leach's, Markham's–White-vented–Wilson's–White-bellied (see below). These gradients, even as viewed now in a relaxed competitive state, involve both the breeding and nonbreeding portions of respective species' populations, and therefore are not necessarily driven by the needs of chick-provisioning parents (see Cody 1974).

Finally, we found an interesting latitudinal pattern of dark versus light color in eastern Pacific storm-petrels (and mirrored in western Pacific forms as well), also apparently in response to habitat gradients. The lightest storm-petrels breed farthest from the tropical–subtropical waters, where the darker species tend to nest. The lightest-colored storm-petrels (e.g., Wilson's and Leach's) are also the most migratory, breeding in subpolar waters but wintering in subtropical and tropical waters.

Thus, an approximate relationship to environmental temperature (Figs. 32–34) is evident, as is the case for color polymorphism in Common Murres (*Uria aalge*; Harris et al. 2003), but in storm-petrels, further study is required to tie the pattern more closely to temperature variation. Common Murres, which nest alongside the European human population, have been intensively studied long enough that in a "natural experiment" caused by changes in ocean temperature during recent decades, changing ratios of color morphs are evident as well.

Coexistence and Competition: Assembly Rules Affecting Storm-petrel Occurrence Patterns

At this point, one might wonder what patterns are evident that are common among the storm-petrel groups in the study area? In other words, we have reviewed the factors that likely led to divergence and speciation, but what are the factors that might be involved if and when species overlap in spatial–temporal occurrence (i.e., are part of the same community)? Are there "rules" that govern whether or not certain storm-petrel species can overlap or co-occur with others, or that prohibit co-occurrence—"assembly rules," as they are called (Diamond 1975)? Of course, teasing out patterns that might suggest rules in this marine avifauna is far more complex than among the terrestrial plant and animal assemblages in which all previous work on this subject has been conducted (reviewed in Weiher and Keddy 1999). As noted above, although storm-petrels may be allopatric in nesting, unlike plants, mammals, and forest birds in that situation, these storm-petrels are often sympatric at sea (or vice versa).

One pattern evident among the storm-petrels is that far more species coexist in the two relatively small eastern boundary currents than in the vast stretches of subtropical and tropical waters that extend toward the equator and to the west in our study area (see Fig. 35): seven species in each current system, if the Fork-tailed Storm-Petrel is also included within the California Current group, but only seven species in the total remaining area, which is 10× more extensive geographically than the current systems (Fig. 1). Primary productivity is extremely high in these two systems (as noted above), compared with the remainder of the study area—which, along with depth-defined microhabitats that affect the composition of potential prey resources, must account for this greater species diversity. Thus, there appears to be a rule whereby overall productivity affects the number of species that can co-occur, at least at the large (current system) scale (see Pianka 1966, Brown 1981). Without further investigation, however, that rule will remain ill-defined for storm-petrels. The fact that only seven storm-petrel species occur along the entire western margin of the Pacific—complex bathymetry as along the eastern margin, and the same latitudinal spread, but much lower productivity —further supports the hypothesis that ocean productivity is involved in explaining species diversity among storm-petrels. A study comparable to the present one but along the margins of the western Pacific is needed to quantify this rule.

Size differentiation, as it affects prey size, has proved to be a primary aspect of community assembly rules in terrestrial environments (Cody 1974; Diamond 1975; Brown 1981; Faaborg 1985, 1988; Brown and Mauer 1987), and this seems to be the case in storm-petrels as well (Fig. 35). However, in storm-petrels, body size wields its effects both on the breeding grounds and on the foraging grounds at the mesoscale or less.

In regard to body size and nesting, storm-petrels of the eastern Pacific Ocean mostly breed on xeric islands where there is little if any soil (including, probably, the habitat islands in the Peruvian coastal desert). Therefore, they make do with whatever nesting cavities they can find. This condition results from the reduced rainfall that characterizes much of the western coast of the midlatitude Americas, the result of the cold upwelling systems offshore. The cold ocean increases the atmospheric pressure in the overlying air masses, which then forces any of the eastward traveling, and wet, low-pressure systems to divert to the north of the southern California Current and to the south of the northern Humboldt Current. In any case, without soil, most storm-petrels in this region do not dig a nesting burrow, as their more polar and subpolar counterparts do, but rather live among the crevices and alleys of soil-less talus slopes.

Ultimately, this condition has led to intense competition for nesting space (recall evidence of "floating populations," above), resolved by body size. In this competition, the situation for storm-petrels is made even more extreme by the suite of even larger non-storm-petrel

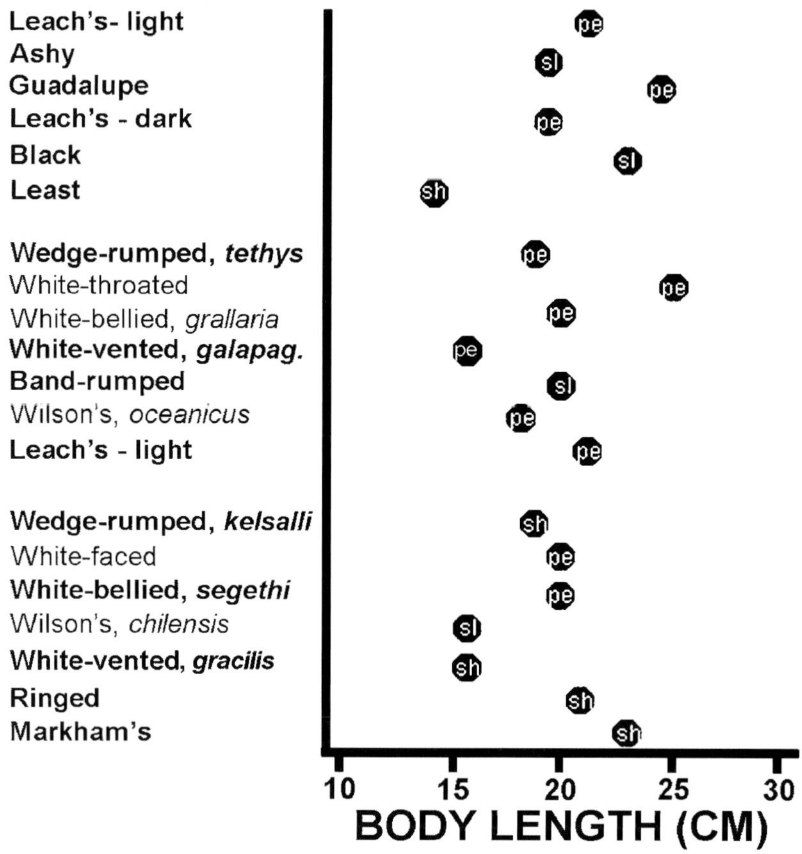

FIG. 35. Sorting of storm-petrel species by body size and mesoscale foraging habitat (sh = shelf, sl = slope, pe = pelagic) in the eastern Pacific Ocean, within the California Current system (top), the Humboldt Current system (bottom), and the vast stretches of subtropical and tropical waters between and to the west (middle). Body size data are from Harrison (1983), adjusted by our own data; mesoscale habitat is from Figures 4 and 19.

species that take the largest cavities (e.g., auklets, murrelets, and larger procellariids). That the competition for nest cavities as a function of body size is very strong is well documented at the Farallon Islands, where smaller and earlier-nesting species or individuals are physically ejected from large cavites by the larger species (Ainley and Boekelheide 1990). The all or mostly dark storm-petrels off Mexico also illustrate this very well, as noted above in the sections on annual cycles (see also Ainley 2005): where these storm-petrel species nest sympatrically, one does not find species of similar size on the same islands (see below). Likely there is some sort of sexual selection going on in relation to body size, but natural selection arising from the reduced availability of nest cavities modifies that (as shown in other vertebrates; e.g., Grant 1986, Wikelski 2005). Cody (1974) argues that the relationship of body size to competition for nesting cavities in seabirds is an artifact secondary to the much more important competition for food at sea, but the storm-petrel patterns argue against this (but see below).

Therefore, one pattern, or perhaps "assembly rule," that stems from the body size factor is that never do more than three storm-petrel taxa nest on the same island. Most commonly, one finds no more than two species co-nesting, and they diverge widely in size or nest in different seasons. Data are too few in regard to the coastal Peruvian desert, but only at three locations—the Galápagos, San Benitos, and Guadalupe islands—are there three co-nesting storm-petrel taxa, all of which diverge widely in size. In the Galápagos, co-occurring species are the White-vented, Wedge-rumped, and Band-rumped storm-petrels, all summer breeders (plus a winter-breeding population of Wedge-rumped Storm-Petrels); at San Benitos, the Black, dark-rumped Leach's, and Least storm-petrels nest in summer; and at Guadalupe, the large Guadalupe Storm-Petrel nests (once nested) with a smaller Leach's Storm-Petrel subspecies, which, in turn, nests out of synchrony with a second, similar-sized Leach's Storm-Petrel subspecies that nests in winter (see Fig. 35).

Not only does body size affect nest-cavity size selection, but it also affects prey size, as in terrestrial systems. Spear et al. (2007) found that among the 30 most abundant seabird species in the eastern tropical Pacific, predator body size has a very strong influence on the size of prey taken. Included within the group of abundant species were six storm-petrel species. Moreover, in the eastern tropical Pacific, a clear gradation exists in diet composition, with the storm-petrels as a group taking small and slow fishes and invertebrates, in contrast to the remaining much larger and faster members of the bird community (Fig. 36). Included in the same trophic guild with storm-petrels are three small gadfly petrels (larger than the storm-petrels): Bulwer's Petrel (*Bulweria bulwerii*), Stejneger's Petrel (*Pterodroma longirostris*), and DeFilippi's Petrel (*P. defilippiana*). Many of the larger, faster seabird species, which are in one of three different foraging guilds, are also among the species that confine

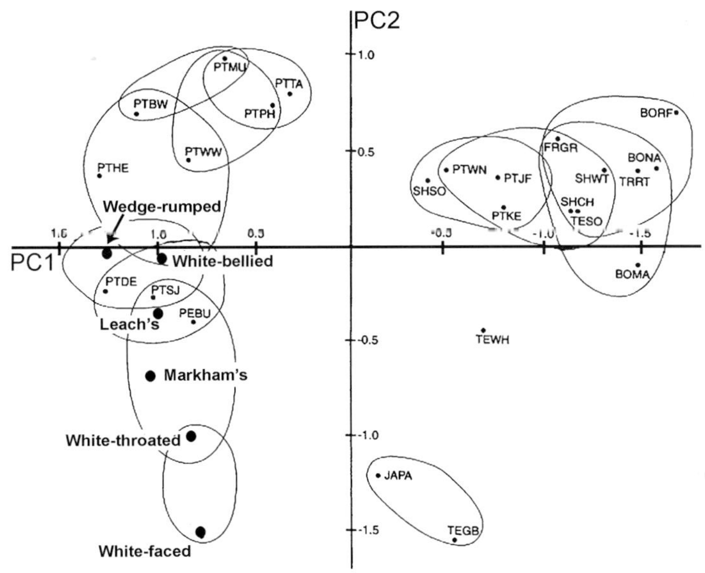

FIG. 36. Results of PCA comparing diet composition among the 30 most abundant species in the slope to pelagic waters of the eastern tropical Pacific Ocean (20°S–20°N), 1983–1991; modified from Spear et al. (2007, which contains the details of diet data collection and analysis). The first and second PC axes explained 45% of the variance in prey species taken by the eastern tropical Pacific avifauna. The most important prey groups on the PC1 axis were myctophids (lantern fish) with positive scores, and the hemirhamphids–exocoetids (flying fish) and epipelagic cephalopods (flying squid) with negative scores. The 15 seabird species that fed predominantly on myctophids were positioned on the positive side, and those that fed on the others were positioned on the negative side. The most important prey groups on the PC2 axis were the negatively loaded miscellaneous invertebrates, and the positively loaded epipelagic cephalopods. Acronyms for non-storm-petrels are as follows: BORF = Red-footed Booby (*Sula sula*), BOMA = Masked Booby (*S. dactylatra*), BONA = Nazca Booby (*S. granti*), FRGR = Great Frigatebird (*Fregata minor*), JAPA = Parasitic Jaeger (*Stercorarius parasiticus*), PEBU = Bulwer's Petrel (*Bulweria bulwerii*), PTBW = Black-winged Petrel (*Pterodroma nigripennis*), PTDE = DeFilippi's Petrel (*P. defilippiana*), PTHE = Herald Petrel (*P. heraldica*), PTJF = Juan Fernandez Petrel (*P. externa*), PTKE = Kermadec Petrel (*P. neglecta*), PTMU = Murphy's Petrel (*P. ultima*), PTPH = Phoenix Petrel (*P. alba*), PTSJ = Stejneger's Petrel (*P. longirostris*), PTTA = Tahiti Petrel (*P. rostrata*), PTWN = White-necked Petrel (*P. cervicalis*), PTWW = White-winged Petrel (*P. leucoptera*), SHCH = Christmas Shearwater (*Puffinus nativitatus*), SHSO = Sooty Shearwater (*P. griseus*), SHWT = Wedge-tailed Shearwater (*P. pacificus*), TEGB = Gray-backed Tern (*Sterna lunata*), TESO = Sooty Tern (*S. fuscata*), TEWH = White Tern (*Gygis alba*), and TRRT = Red-tailed Tropicbird (*Phaethon rubricauda*).

storm-petrels to the use of the smallest nesting cavities (see above).

Among those six storm-petrel species for which there were adequate samples in the eastern tropical Pacific, in turn, it is evident that there is a clear sorting by diet composition, ranging from mostly fish taken by the Wedge-rumped Storm-Petrel to mostly invertebrates by the White-faced Storm-Petrel (Fig. 37). Likely, aspects of flight behavior and speed probably facilitate this divergence. Thus, unlike terrestrial vertebrates studied for assembly rules, where diet classes are much more clearly defined (e.g., insectivore, herbivore, gramnivore; Diamond 1975, Faaborg 1988, Fox 1999), one cannot classify most storm-petrels so easily. Where there is strong diet overlap, other factors come into play to allow coexistence, such as a temporal aspect to foraging. For example, among the co-occurring Wedge-rumped and Leach's storm-petrels, the former forages mostly at night and the latter mostly during the day or in crepuscular hours.

Finally and in sum, Lack (1971) suggested that nearly all avian species separate by some combination of range, habitat, food, and, in some cases, time. Apparently, application of just two or three factors explains the ecological isolation of most coexisting storm-petrels, allowing for more overlap than in terrestrial systems, because these birds wander across large stretches of ocean between sometimes disconnected breeding-, foraging-, and nonbreeding-season habitat. Where and when storm-petrels

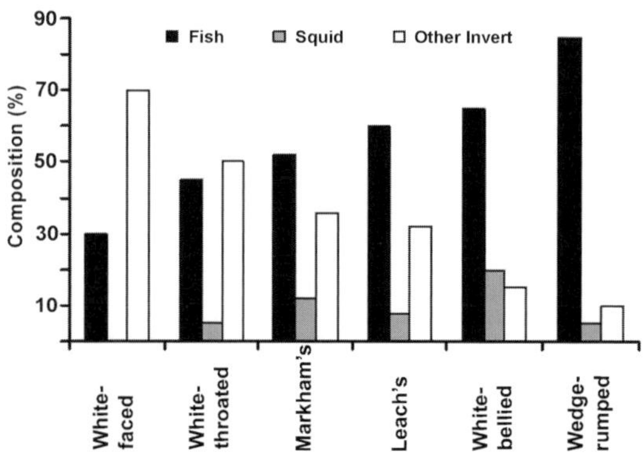

FIG. 37. Diet composition among three major prey groups for the six most abundant storm-petrel species in the slope to pelagic waters of the eastern tropical Pacific Ocean (20°S–20°N). Data are summarized from Spear et al. (2007; see that source for information on methods of diet data collection and analysis).

co-occur (see again Fig. 35), they have to be of markedly different size, especially on the nesting grounds. At sea, if they do not diverge in size (thus to take different-sized prey), they diverge in foraging microhabitat, sorting along the shelf–slope–pelagic gradient. These two factors explain co-occurrence for most storm-petrel species. Mechanisms related to foraging behavior further separate the few remaining co-occurring species. Similarly, Hunter (1987) noted that co-occurrence is possible between the two sibling Giant Petrel species (*Macronectes giganteus* and *M. halli*), in which there is no difference in body size, by virtue of differences in diet and breeding phenology. Therefore, with these Giant Petrels behaving like the two sibling subspecies of *O. leucorhoa* at Guadalupe Island, it seems that the few rules that govern co-occurrence of storm-petrels may operate among other seabird species assemblages as well.

Further Research Needed

Obviously, a study of the molecular taxonomy of this family in the eastern Pacific would be very interesting, given the general isolation of many breeding populations, the diversity of the taxa involved, and the information reported here. Secondly, information on diet of the eastern boundary current species would allow a better understanding of mechanisms of co-occurrence or divergence; currently, we know about diet only for the eastern tropical Pacific species. Third, surveys of nesting areas in coastal Peru as well as in the Austral Islands will provide much-needed information on storm-petrel status. And, finally, a comparable study of storm-petrels of the western Pacific (or perhaps in areas of the Atlantic) would help us to better understand how ocean productivity affects the mesoscale division of ocean habitat that is evident in this seabird family.

Acknowledgments

We thank the officers and crews of the National Oceanic and Atmospheric Administration (NOAA) vessels *David Starr Jordan, Discoverer, Malcolm Baldrige, Oceanographer*, and *Surveyor*, and of the National Science Foundation's (NSF) *R/V Hero*. R. Kelton and the Kelton Foundation provided a boat for (and sponsored) two additional cruises off Mexico. Much of this study was done in conjunction with the Tropical Ocean–Global Atmosphere and Equatorial Pacific Ocean Climate

Studies and rockfish assessment cruises of NOAA. Participation on NOAA cruises was made possible by Pacific Marine Environmental Laboratories, Atlantic Marine Oceanographic Laboratories, and the Soutwest Fisheries Science Center. Field assistance by I. Gaffney, K. Hansen, P. Henderson, S. Howell, N. Karnovsky, L. Logerwell, P. Pyle, C. Owston, M. Rauzon, C. Strong, T. Wahl, and S. Webb was invaluable, as was that of K. Warheit who reviewed portions of the manuscript. The guidance by J. Faaborg through several revisions of the manuscript was invaluable, and S. Howell kindly allowed the use of his photos. *Hero* and NOAA cruises were funded by NSF grants DPP7820755, OCE8515637, and OCE8911125 and National Geographic Society grants 3321-86 and 4106-89. An important portion of the writing was completed under NSF grant OCE 0534609, and this paper is contribution number 331 of Northeast Pacific GLOBEC. Ocean depth and distance to mainland were calculated using coastline and bathymetry data obtained from the National Geophysical Data Center (rimmer.ngdc.noaa.gov/mgg/coast/getcoast.html) and WORLDBATH (ingrid.ldgo.columbia.edu/SOURCES/WORLDBATH), respectively.

Literature Cited

Ainley, D. G. 1976. The occurrence of seabirds in the coastal region of California. Western Birds 7:33–68.

Ainley, D. G. 1977. Feeding methods in seabirds: A comparison of polar and tropical communities in the eastern Pacific Ocean. Pages 669–685 *in* Adaptations within Antarctic Ecosystems (G. A. Llano, Ed.). Gulf Publishing, Houston, Texas.

Ainley, D. G. 1980. Geographic variation in the Leach's Storm-Petrel. Auk 97:837–853.

Ainley, D. G. 1995. Ashy Storm-Petrel (*Oceanodroma homochroa*). *In* The Birds of North America, no. 185 (A. Poole and F. Gill, Eds.). Academy of Natural Sciences, Philadelphia, and American Ornithologists' Union, Washington, D.C.

Ainley, D. G. 2005. The dark storm-petrels of the eastern Pacific: Speciation, current status, and future prospects. Birding (February):58–65.

Ainley, D. G., and R. J. Boekelheide, Eds. 1990. Seabirds of the Farallon Islands: Ecology, Dynamics, and Structure of an Upwelling-system Community. Stanford University Press, Stanford, California.

Ainley, D. G., and W. T. Everett. 2001. Black Storm-Petrel (*Oceanodroma melania*). *In* The Birds of North America, no. 577 (A. Poole and F. Gill, Eds.). Birds of North America, Philadelphia.

Ainley, D. G., R. P. Henderson, and C. S. Strong. 1990. Leach's Storm-Petrel and Ashy Storm-Petrel. Pages 128–162 *in* Seabirds of the Farallon Islands (D. G. Ainley and R. J. Boekelheide, Eds.). Stanford University Press, Stanford, California.

Ainley, D. G., T. J. Lewis, and S. Morrell. 1976. Molt in Leach's and Ashy storm-petrels. Wilson Bulletin 88:76–95.

Ambagis, J. 2004. A comparison of census and monitoring techniques for Leach's Storm-Petrel. Waterbirds 27:211–215.

Astheimer, L. B., and C. R. Grau. 1990. A comparison of yolk growth rates in seabird eggs. Ibis 132:380–394.

Austin, O. L., Jr. 1952. Notes on some petrels of the North Pacific. Bulletin of the Harvard Museum of Comparative Zoology 107:391–407.

Baker, G. B., R. Gales, S. Hamilton, and V. Wilkinson. 2002. Albatrosses and petrels in Australia: A review of their conservation and management. Emu 102:71–97.

Ballance, L. T., R. L. Pitman, and S. B. Reilly. 1997. Seabird community structure along a productivity gradient: Importance of competition and energetic constraint. Ecology 78: 1502–1518.

Bell, B., P. Fullagar, C. Robertson, and J. Van Tets. 1984. Status of Australian seabirds. Australian Seabird Group Newsletter 20:1–24.

Boersma, P. D., and M. C. Silva. 2001. Fork-tailed Storm-Petrel (*Oceanodroma furcata*). *In* The Birds of North America, no. 569 (A. Poole and F. Gill, Eds.). Birds of North America, Philadelphia.

Briggs, K. T., W. B. Tyler, D. B. Lewis, and D. R. Carlson. 1987. Bird communities at sea off California: 1975 to 1983. Studies in Avian Biology, no. 11.

Brooke, M. 2004. Albatrosses and Petrels across the World. Oxford University Press, Oxford, United Kingdom.

Brooke, M. de L. 1987. The birds of the Juan Fernandez Islands, Chile. Study Report no. 16, International Council for Bird Preservation, Cambridge, United Kingdom.

Broughton, J. M. 1994. Size of the bursa of Fabricius in relation to gonad size and age in Laysan and Black-footed albatrosses. Condor 96:203–207.

Brown, J. H. 1981. Two decades of homage to Santa Rosalia: Toward a general theory of diversity. American Zoologist 21:877–888.

Brown, J. H., and B. A. Maurer. 1987. Evolution of species assemblages: Effects of energetic constraints and species dynamics on the diversification of the North American avifauna. American Naturalist 130:1–17.

Campos, A. R., and J. P. Granadeiro. 1999. Breeding biology of the White-faced Storm-Petrel on Selvagem Grande Island, north-east Atlantic. Waterbirds 22:199–206.

Cheke, R. A. 1967. Observations on Japanese seabirds. Sea Swallow 19:31–35.

Clarke, E. D., L. B. Spear, M. L. McCracken, F. F. C. Marques, D. L. Borchers, S. T. Buckland, and D. G. Ainley. 2003. Validating the use of generalized additive models and at-sea surveys to estimate size and temporal trends of seabird populations. Journal of Applied Ecology 40: 278–292.

Cody, M. L. 1974. Competition and the Structure of Bird Communities. Princeton University Press, Princeton, New Jersey.

Coulter, M. C. 1984. Seabird conservation in the Galápagos Islands, Ecuador. Pages 237–244 *in* Status and Conservation of the World's Seabirds (J. P. Croxall, P. G. H. Evans, and R. W. Schreiber, Eds.). Technical Publication no. 2, International Council for Bird Preservation, Cambridge, United Kingdom.

Crossin, R. S. 1974. The storm-petrels (Hydrobatidae). Pages 154–205 *in* Pelagic Studies of Seabirds in the central and eastern Pacific Ocean (W. B. King, Ed.). Smithsonian Contributions to Zoology, no. 158.

Croxall, J. P., P. A. Prince, I. Hunter, S. J. McInnes, and P. G. Copestake. 1984. The seabirds of the Antarctic Peninsula, islands of the Scotia Sea and Antarctic continent between 80°W and 20°W. Pages 637–666 *in* Status and Conservation of the World's Seabirds (J. P. Croxall, P. G. H. Evans, and R. W. Schreiber, Eds.). Technical Publication no. 2, International Council for Bird Preservation, Cambridge, United Kingdom.

Davison, A. C., and D. V. Hinkley. 1997. Bootstrap Methods and Their Application. Cambridge University Press, London.

Diamond, J. M. 1975. Assembly of species communities. Pages 342–444 *in* Ecology and Evolution of Communities (M. L. Cody and J. M. Diamond, Eds.). Belknap Press of Harvard University Press, Cambridge, Massachusetts.

Duffy, D. C., C. Hays, and M. Plenge. 1984. The status and conservation of Peruvian seabirds. Pages 245–259 *in* Status and Conservation of the World's Seabirds (J. P. Croxall, P. G. H. Evans, and R. W. Schreiber, Eds.). Technical Publication no. 2, International Council for Bird Preservation, Cambridge, United Kingdom.

Efron, B., and R. J. Tibshirani. 1993. An Introduction to the Bootstrap. Chapman and Hall, London.

Enticott, J., and P. Tipling. 1997. Photographic Handbook of the Seabirds of the World. New Holland, London.

Everett, W. T., and D. W. Anderson. 1991. Status and conservation of the breeding seabirds on onshore Pacific Islands of Baja California and the Gulf of California. Pages 115–139 *in* Seabird Status and Conservation: A Supplement (J. P. Croxall, Ed.). Technical Publication no. 11, International Council for Bird Preservation, Cambridge, United Kingdom.

Faaborg, J. 1985. Ecological constraints on West Indian bird distributions. Pages 621–653 *in* Neotropical Ornithology (P. A. Buckley, M. S. Foster, E. S. Morton, R. S. Ridgely, and F. G. Buckley, Eds.). Ornithological Monographs, no. 36.

Faaborg, J. 1988. Ornithology: An Ecological Approach. Prentice Hall, Englewood Cliffs, New Jersey.

Falla, R. A. 1933. Notes on New Zealand petrels, with descriptions of new forms and some new records. Records of the Auckland Institute Museum 1:173–180.

Fiedler, P. C., V. Philbrick, and F. P. Chavez. 1991. Oceanic upwelling and productivity in the eastern tropical Pacific. Limnology and Oceanography 36:1834–1850.

Fox, B. J. 1999. The genesis and development of guild assembly rules for guilds. Pages 23–57 *in* Ecological Assembly Rules: Perspectives, Advances, Retreats (E. Weiher and P. Keddy, Eds.). Cambridge University Press, Cambridge, United Kingdom.

Fullagar, P. J., J. L. McKean, and G. F. Van Tets. 1974. Appendix F. Report on the birds. Pages 55–72 *in* Environmental Survey of Lord Howe Island (H. F. Recher and S. S. Clark, Eds.). New South Wales Government Printer, Sydney, Australia.

Garcia-Paris, M., D. A. Good, G. Parra-Olea, and D. B. Wake. 2000. Biodiversity of Costa Rican salamanders: Implications of high levels of genetic differentiation and phylogeographic structure for species formation. Proceedings of the National Academy of Sciences USA 97: 1640–1647.

Garnett, M. C. 1984. Conservation of seabird in the South Pacific region: A review. Pages 547–558 *in* Status and Conservation of the World's Seabirds (J. P. Croxall, P. G. H. Evans, and R. W. Schreiber, Eds.). Technical Publication no. 2, International Council for Bird Preservation, Cambridge, United Kingdom.

Genovart, M., D. Oro, and F. Bonhomme. 2003. Genetic and morphological differentiation between the two largest breeding colonies of Audouin's Gull *Larus audouinii.* Ibis 145: 448–456.

Glantz, M. H., and J. D. Thompson, Eds. 1981. Resource Management and Environmental Uncertainty: Lessons from Coastal Upwelling Fisheries. John Wiley and Sons, New York.

GRANT, P. R. 1986. Ecology and Evolution of Darwin's Finches. Princeton University Press, Princeton, New Jersey.

HARRIS, M. P. 1969. The biology of storm-petrels in the Galápagos Islands. Proceedings of the California Academy of Sciences 37:95–166.

HARRIS, M. P., P. ROTHERY, AND S. WANLESS. 2003. Increase in frequency of the bridled morph of the Common Guillemot *Uria aalge* on the Isle of May, 1946–2000: A return to former levels? Ibis 145:22–29.

HARRISON, C. S., M. B. NAUGHTON, AND S. I. FEFER. 1984. The status and conservation of seabirds in the Hawaiian Archipelago and Johnston Atoll. Pages 513–526 *in* Status and Conservation of the World's Seabirds (J. P. Croxall, P. G. H. Evans, and R. W. Schreiber, Eds.). Technical Publication no. 2, International Council for Bird Preservation, Cambridge, United Kingdom.

HARRISON, P. 1983. Seabirds: An Identification Guide. Houghton Mifflin, Boston, Massachusetts.

HASEGAWA, H. 1984. Status and conservation of seabirds in Japan, with special attention to the Short-tailed Albatross. Pages 487–500 *in* Status and Conservation of the World's Seabirds (J. P. Croxall, P. G. H. Evans, and R. W. Schreiber, Eds.). Technical Publication no. 2, International Council for Bird Preservation, Cambridge, United Kingdom.

HASTIE, T., AND R. J. TIBSHIRANI. 1990. Generalized Additive Models. Chapman and Hall, London.

HEINEMANN, D. 1981. A range finder for pelagic bird censusing. Journal of Wildlife Management 45: 489–493.

HOLYOAK, D. T., AND J.-C. THIBAULT. 1984. Contribution à l'étude des oiseaux de Polynésie Orientale. Memoires of the Museum Natural History Service A 127:1–209.

HOWELL, S. N. G., AND S. WEBB. 1995. A Guide to the Birds of Mexico and Northern Central America. Oxford University Press, Oxford, United Kingdom.

HUBER, L. N. 1971. Notes on the migration of the Wilson's Storm-Petrel (*Oceanites oceanicus*) near Eniwetok Atoll, western Pacific Ocean. Notornis 18:38–42.

HUEY, L. M. 1952. *Oceanodroma tethys tethys*, a petrel new to the North American avifauna. Auk 69: 460–461.

HUNT, G. L., JR. 1990. The pelagic distribution of marine birds in a heterogeneous environment. Polar Research 8:43–54.

HUNTER, S. 1987. Species and sexual isolating mechanisms in sibling species of giant petrels *Macronectes*. Polar Biology 7:295–301.

HUNTINGTON, C. E., R. G. BUTLER, AND R. A. MAUCK. 1996. Leach's Storm-Petrel (*Oceanodroma leucorhoa*). *In* The Birds of North America, no. 233 (A. Poole and F. Gill, Eds.). Academy of Natural Sciences, Philadelphia, and American Ornithologists' Union, Washington, D.C.

IMBER, M. J. 1984. Migration of White-faced Storm-Petrels *Pelagodroma marina* in the South Pacific and the status of the Kermadec subspecies. Emu 84:32–35.

IMBER, M. J., AND T. G. LOVEGROVE. 1982. Leach's Storm Petrels (*Oceanodroma l. leucorhoa*) prospecting for nest sites on the Chatham Islands. Notornis 29:101–108.

JAHNCKE, J. 1993. First report on the breeding areas of the Markham's Storm-Petrel *Oceanodroma markhami* (Salvin, 1883). Pages 339–343 *in* Memoria X Congreso Nacional de Biologia (E. Castillo de Maruenda, Ed.). Lima, Perú.

JEHL, J. R., JR. 1984. Conservation problems of seabirds in Baja California and the Pacific Northwest. Pages 41–48 *in* Status and Conservation of the World's Seabirds (J. P. Croxall, P. G. H. Evans, and R. W. Schreiber, Eds.). Technical Publication no. 2, International Council for Bird Preservation, Cambridge, United Kingdom.

KLEINBAUM, D. G., L. L. KUPPER, AND K. E. MULLER. 1988. Applied Regression Analysis and Other Multivariable Methods. PWS-KENT, Boston, Massachusetts.

KLOMP, N. I., AND R. W. FURNESS. 1992. Non-breeders as a buffer against environmental stress: Declines in numbers of Great Skuas on Foula, Shetland, and prediction for future recruitment. Journal of Applied Ecology 29: 341–348.

LACK, D. 1971. Ecological Isolation in Birds. Blackwell, London.

LENSINK, C. J. 1984. The status and conservation of seabird in Alaska. Pages 13–27 *in* Status and Conservation of the World's Seabirds (J. P. Croxall, P. G. H. Evans, and R. W. Schreiber, Eds.). Technical Publication no. 2, International Council for Bird Preservation, Cambridge, United Kingdom.

LÉVÊQUE, R., R. I. BOWMAN, AND S. L. BILLEB. 1966. Migrants in the Galápagos area. Condor 68: 81–101.

LOCKLEY, R. M. 1983. Flight of the Storm Petrel. Paul S. Eriksson, Middlebury, Vermont.

LONGHURST, A. R., AND D. PAULY. 1987. Ecology of Tropical Oceans. Academic Press, San Diego, California.

LOOMIS, L. M. 1918. A review of the albatrosses, petrels, and diving petrels. Proceedings of the California Academy of Sciences, 4th Series, vol. 2.

MANUWAL, D. A. 1974. Effects of territoriality on breeding in a population of Cassin's Auklet. Ecology 55:1399–1406.

Marchant, S., and P. J. Higgins, Eds. 1990. Handbook of Australian, New Zealand, and Antarctic Birds, vol. 1: Ratites to Ducks. Oxford University Press, Melbourne, Australia.

Mayr, E. 1964. Systematics and the Origin of Species, from the Viewpoint of a Zoologist. Dover, New York.

McCaskie, G. 1990. First record of the Band-rumped Storm-Petrel in California. Western Birds 21:65–68.

McCullagh, P., and J. A. Nelder. 1989. Generalized Linear Models, 2nd ed. Chapman and Hall, London.

Moffitt, J. 1938. Two southern petrels in the North Pacific. Auk 55:255–259.

Murphy, R. C. 1924. Birds collected during the Whitney South Sea Expedition. II. American Museum Novitates, no. 124.

Murphy, R. C. 1936. Oceanic Birds of South America. Macmillan, New York.

Murphy, R. C., and S. Irving. 1951. A review of the frigate-petrels (*Pelagodroma*). American Museum Novitates, no. 1506.

Murphy, R. C., and J. P. Snyder. 1952. The "Pealea" phenomenon and other notes on storm petrels. American Museum Novitates, no. 1596.

Nagy, K. A. 1987. Field metabolic rate and food requirements scaling in mammals and birds. Ecological Monographs 57:111–128.

Oedekoven, C. S., D. G. Ainley, and L. B. Spear. 2001. Variable responses of seabirds to change in marine climate: California Current, 1985–1994. Marine Ecology Progress Series 212: 265–281.

Paulik, G. J. 1981. Anchovies, birds, and fishermen in the Peru Current. Pages 35–79 *in* Resource Management and Environmental Uncertainty: Lessons from Coastal Upwelling Fisheries (M. H. Glantz and J. D. Thompson, Eds.). John Wiley and Sons, New York.

Pennycuick, C. J. 1989. Bird Flight Performance: A Practical Calculation Manual. Oxford University Press, Oxford, United Kingdom.

Pianka, E. R. 1966. Latitudinal gradients in species diversity: A review of concepts. American Naturalist 100:33–46.

Power, D. M., and D. G. Ainley. 1986. Seabird geographic variation: Similarity among populations of Leach's Storm-Petrel. Auk 103: 575–585.

Pyle, P. 1993. A Markham's Storm-Petrel in the northeastern Pacific. Western Birds 24:108–110.

Ribic, C. A., and D. G. Ainley. 1988. Constancy of seabird assemblages: An exploratory look. Biological Oceanography 6:175–202.

Ribic, C. A., and D. G. Ainley. 1997. The relationships of seabird assemblages to physical habitat features in Pacific equatorial waters during spring 1984–1991. ICES Journal of Marine Science 54: 593–599.

Ribic, C. A., D. G. Ainley, and L. B. Spear. 1997. Scale-related seabird–environmental relationships in Pacific equatorial waters, with reference to El Niño–Southern Oscillation events. Marine Ecology Progress Series 156:183–203.

Robertson, C. J. R., and B. D. Bell. 1984. Seabird status and conservation in the New Zealand region. Pages 573–586 *in* Status and Conservation of the World's Seabirds (J. P. Croxall, P. G. H. Evans, and R. W. Schreiber, Eds.). Technical Publication no. 2, International Council for Bird Preservation, Cambridge, United Kingdom.

Rundle, H. D., and D. Schluter. 2004. Natural selection and ecological speciation in sticklebacks. Pages 192–209 *in* Adaptive Speciation (U. Dieckmann, M. Doebeli, J. A. J. Metz, and D. Tautz, Eds.). Cambridge University Press, Cambridge, United Kingdom.

Ryan, P. G., W. R. J. Dean, C. L. Moloney, B. P. Watkins, and S. J. Milton. 1990. New information on seabirds at Inaccessible Island and other islands in the Tristan da Cunha group. Marine Ornithology 18:43–54.

SAS Institute. 1985. SAS User's Guide: Statistics, 5th ed. SAS Institute, Cary, North Carolina.

Schlatter, R. P. 1984. The status and conservation of seabirds in Chile. Pages 261–269 *in* Status and Conservation of the World's Seabirds (J. P. Croxall, P. G. H. Evans, and R. W. Schreiber, Eds.). Technical Publication no. 2, International Council for Bird Preservation, Cambridge, United Kingdom.

Schlatter, R. P., and M. A. Marin. 1983. Breeding of Elliot's Storm-Petrel, *Oceanites gracilis*, in Chile. Gerfaut 73:197–199.

Schreiber, R. W., and N. P. Ashmole. 1970. Seabird breeding seasons on Christmas Island, Pacific Ocean. Ibis 112:363–394.

Seber, G. A. F. 1977. Linear Regression Analysis. John Wiley and Sons, New York.

Snow, D. W., and C. M. Perrins, Eds. 1998. The Birds of the Western Palearctic, vol. 1. Oxford University Press, Oxford, United Kingdom.

Spear, L. B. 2001. Birds: Seabird migration. Pages 2669–2677 *in* Encyclopedia of Ocean Sciences (J. Steele, S. Thorpe, and K. Tarekian, Eds.). Academic Press, London.

Spear, L. B., and D. G. Ainley. 1997a. Flight speed of seabirds in relation to wind speed and direction. Ibis 139:234–251.

Spear, L. B., and D. G. Ainley. 1997b. Flight behaviour of seabirds in relation to wind direction and wing morphology. Ibis 139:221–233.

Spear, L. B., and D. G. Ainley. 1998. Morphological differences relative to ecological segregation in petrels (Family: Procellariidae) of the Southern Ocean and tropical Pacific. Auk 115:1017–1033.

Spear, L. B., and D. G. Ainley. 1999. Seabirds of the Panama Bight. Waterbirds 22:175–198.

Spear, L. B., and D. G. Ainley. 2005a. At-sea distributions and abundance of tropicbirds in the eastern Pacific. Ibis 147:353–366.

Spear, L. B., and D. G. Ainley. 2005b. At-sea behaviour and habitat use by tropicbirds in the eastern Pacific. Ibis 147:391–407

Spear, L. B., D. G. Ainley, B. D. Hardesty, S. N. G. Howell, and S. W. Webb. 2004. Reducing biases affecting at-sea surveys of seabirds: Use of multiple observer teams. Marine Ornithology 32:147–157.

Spear, L. B., D. G. Ainley, N. Nur, and S. N. G. Howell. 1995a. Population size and factors affecting the at-sea distributions of four endangered procellariids in the tropical Pacific. Condor 97:613–638.

Spear, L. B., D. G. Ainley, and C. A. Ribic. 1995b. Incidence of plastic in seabirds from the tropical Pacific, 1984–1991: Relation with distribution of species, sex, age, season, year and body weight. Marine Environmental Research 40:123–146.

Spear, L. B., D. G. Ainley, and W. A. Walker. 2007. Diet and foraging dynamics of pelagic seabirds in the eastern Pacific Ocean. Studies in Avian Biology. In press.

Spear, L. B., D. G. Ainley, and S. W. Webb. 2003. Distribution, abundance, and behaviour of Buller's, Chatham Island and Salvin's albatrosses off Chile and Peru. Ibis 145:253–269.

Spear, L. B., D. G. Ainley, and S. W. Webb. 2005. Distribution, abundance, habitat use and behavior of three *Procellaria* petrels off South America. Notornis 52:88–105.

Spear, L. B., L. T. Ballance, and D. G. Ainley. 2001. Response of seabirds to thermal boundaries in the tropical Pacific: The thermocline vs. the Equatorial Front. Marine Ecology Progress Series 219:275–289.

Spear, L. B., N. Nur, and D. G. Ainley. 1992. Estimating absolute densities of flying seabirds using analyses of relative movement. Auk 109:385–389.

Sydeman, W. J., N. Nur, E. B. McLaren, and G. J. McChesney. 1998. Status and trends of the Ashy Storm-Petrel on Southeast Farallon Island, California, based upon capture–recapture analyses. Condor 100:438–447.

Taylor, G. A. 2000. Action plan for seabird recovery in New Zealand. Biodiversity Recovery Unit, Department of Conservation, Wellington, New Zealand.

Tennyson, A., and G. Taylor. 1990. Curtis Island. Ornithological Society of New Zealand News 57:10.

Thibault, J.-C. 1973. Notes ornithologiques Polynésiennes II: Les Iles Marquises. Alauda 41:301–316.

Trenberth, K. E. 1997. The definition of El Niño. Journal of the American Meteorological Society 78:2771–2777.

Van Tets, G. F., and P. J. Fullagar. 1984. Status of seabirds in Australia. Pages 559–571 *in* Status and Conservation of the World's Seabirds (J. P. Croxall, P. G. H. Evans, and R. W. Schreiber, Eds.). Technical Publication no. 2, International Council for Bird Preservation, Cambridge, United Kingdom.

Velarde, E., and D. W. Anderson. 1994. Conservation and management of seabird islands in the Gulf of California: Setbacks and successes. Pages 229–243 *in* Seabirds on Islands: Threats, Case Studies and Action Plans (D. N. Nettleship, J. Burger, and M. Gochfeld, Eds.). Birdlife International, Cambridge, United Kingdom.

Venables, W. N., and B. D. Ripley. 1997. Modern Applied Statistics with S-PLUS, 2nd ed. Springer, New York.

Vilchis, L. I., L. T. Ballance, and P. C. Fiedler. 2006. Pelagic habitat of seabirds in the eastern tropical Pacific: Effect of foraging ecology on habitat selection. Marine Ecology Progress Series 315:279–292.

Warham, J. 1990. The Petrels: Their Ecology and Breeding Systems. Academic Press, London.

Warham, J. 1996. The Behaviour, Population Biology and Physiology of the Petrels. Academic Press, London.

Warheit, K. I. 1992. A review of the fossil seabirds of the Tertiary of the North Pacific: Plate tectonics, paleoceanography, and faunal change. Paleobiology 18:401–424.

Warheit, K. I. 2002. The seabird fossil record and the role of paleontology in understanding seabird community structure. Pages 17–56 *in* Biology of Marine Birds (E. A. Schreiber and J. Burger, Eds.). CRC Publications, Boca Raton, Florida.

Weiher, E., and P. Keddy, Eds. 1999. Ecological Assembly Rules: Perspectives, Advances, Retreats. Cambridge University Press, Cambridge, United Kingdom.

Weimerskirch, H. 1991. Sex-specific differences in molt strategy in relation to breeding in the Wandering Albatross. Condor 93:731–737.

Wikelski, M. 2005. Evolution of body size in Galápagos marine iguanas. Proceedings of the Royal Society of London, Series B 272:1985–1993.

Wyrtki, K. 1966. Oceanography of the eastern equatorial Pacific Ocean. Oceanography and Marine Biology 4:33–68.

Appendix 1. Measurements of Wilson's Storm-Petrels (*O. o. oceanities*, *O. o. chilensis*) and White-vented Storm-Petrels (*O. g. gracilis*). Values are means, with ranges in parentheses.

Mass (g)	Chord (mm)	Tail (mm)	Culmen (mm)	Tarsus (mm)
(A) *O. o. oceanities*; South Shetland Islands, Antarctic Peninsula (*n* = 17; Murphy 1936).				
–	149.8	67.6	12.9	34.7
	(144–158)	(63–73)	(12.1–13.8)	(33.6–36.5)
(B) *O. o. oceanities*; eastern Pacific (*n* = 6; present study).				
36.0	153.7	69.7	12.8	35.8
(34–38)	(149–162)	(63–74)	(11.9–13.5)	(33.5–38.0)
(C) *O. o. chilensis*; on or near colony, Strait of Magellan, Chile (*n* = 11; Murphy 1936).				
–	136.1	59.5	11.3	35.2
(D) *O. o. chilensis*; 75 km off Ecuador (*n* = 7; present study).				
30.7	137.9	61.3	11.3	35.3
(28–34)	(134–148)	(57–65)	(10.5–12.0)	(33.7–36.9)
(E) *O. g. gracilis*; collected off Peru and Chile (*n* = 46; Murphy 1936).				
–	125.8	56.0	11.1	30.0
	(117–132)	(48–57)	(10.5–11.6)	(28.0–31.5)
(F) *O. g. gracilis*; 75 km off Ecuador (*n* = 2; present study).				
16.5	123.5	52.5	10.8	29.1
(16–17)	(118–129)	(50–55)	(10.6–10.9)	(28.3–29.9)

APPENDIX 2. Measurements of White-bellied Storm-Petrels. Birds were collected (A) at Rapa Island breeding colony; (B, I) in the eastern equatorial Pacific; (C) in the southern Atlantic and Indian oceans; (D) at Tristan da Cunha; (E, F, G) in the breeding colony at Lord Howe Island; (H) near Marquesas; (J) near the breeding colony at Juan Fernandez Islands, Chile; and (K) in the Humboldt Current. Known or most likely taxa are also given; taxa collected in the present study are in bold. Values are means, with ranges in parentheses.

Mass (g)	Chord (mm)	Tail (mm)	Culmen (mm)	Tarsus (mm)
(A) *F. g. titan*; Rapa Island, Austral Islands (*n* = 27; Murphy and Snyder 1952).				
—	182.0	83.1	15.5	41.1
	(177–188)	(78.0–89.0)	(15.0–16.3)	(39.0–43.0)
(B) ***F. g. titan***; eastern equatorial Pacific (*n* = 1; this study).				
64.0	181.0	84.0	15.9	39.3
(C) *F. g. leucogaster*; southern Atlantic and Indian oceans (*n* = 6; Murphy and Snyder 1952).				
—	167.5	74.0	14.6	38.9
	(160–172)	(69.7–79.8)	(13.0–15.1)	(36.0–41.0)
(D) *F. g. leucogaster*; Tristan da Cunha (*n* = 7; Murphy 1936).				
—	162.5	77.2	14.6	38.9
	(156–170)	(71–82)	(14.0–15.6)	(36.0–42.3)
(E) *F. g. grallaria*; Lord Howe Island breeding population (skins; *n* = 4; Murphy and Snyder 1952).				
—	163.5	77.2	13.9	35.8
	(161–166)	(75–78)	(13.5–14.1)	(34.8–37.1)
(F) *F. g. grallaria*; Lord Howe Island (skins; *n* = 8; Marchant and Higgins 1990).				
165	75.0	1	4	37.6
	(161–169)	(71–78)	(13.0–14.7)	(35.6–39.6)
(G) *F. g. grallaria*; Lord Howe Island (field measurements; *n* = 37; Marchant and Higgins 1990).				
52.0	168.0	75.6	14.1	37.1
(45– 65)	(159–176)	(66–82)	13.2–15.0)	(35.1–39.1)
(H) *F. g. lineata*; Marquesas (*n* = 1; Murphy and Snyder 1952).				
—	165.0	73.5	14.0	38.0
(I) ***F. g. grallaria***; eastern equatorial Pacific from 120°W to 150°W (*n* = 15; present study).				
47.4	172.0	79.3	14.5	37.3
(40–62)	(164–180)	(77–83)	(14.2–14.6)	(34.7–39.1)
(J) *F. g. segethi*; Juan Fernandez Islands breeding colony (*n* = 63; Murphy and Snyder 1952, Murphy 1936).				
—	155.9	73.4	13.3	35.3
	(146–163)	(71.0–77.0)	(12.6–14.0)	(33.0–37.0)
(K) ***F. g. segethi***; Humboldt Current and eastern Pacific from 83°W to 110°W (*n* = 7; present study).				
38.9	165.5	75.0	14.5	37.0
(34–43)	(161–171)	(71.0–79.0)	(14.5–14.6)	(35.0–39.0)